Contributions to Economics

The series *Contributions to Economics* provides an outlet for innovative research in all areas of economics. Books published in the series are primarily monographs and multiple author works that present new research results on a clearly defined topic, but contributed volumes and conference proceedings are also considered. All books are published in print and ebook and disseminated and promoted globally. The series and the volumes published in it are indexed by Scopus and ISI (selected volumes).

Julia Puaschunder

Advances in Behavioral Economics and Finance Leadership

Strategic Leadership, Wise Followership and Conscientious Usership in the Digital Century

Second Edition

Springer

Julia Puaschunder
Graduate School of Arts and Sciences
Columbia University
New York, NY, USA

ISSN 1431-1933 ISSN 2197-7178 (electronic)
Contributions to Economics
ISBN 978-3-031-15709-7 ISBN 978-3-031-15710-3 (eBook)
https://doi.org/10.1007/978-3-031-15710-3

This Springer imprint is published by the registered company Springer Nature Switzerland AG
The registered company address is: Gewerbestrasse 11, 6330 Cham, Switzerland

Preface

Behavioral economics revolutionized decision-making theory. Behavioral insights offer promising avenues to help people make better and more socially responsible choices in very many different domains. The behavioral revolution changed economic theory and policy practice in fields ranging from marketing, corporate governance to public affairs as well as leadership. Behavioral economists nowadays nudge and wink people into favorable decision outcomes.

This book aims at nurturing interdisciplinary interests in behavioral economics in the digital century. Behavioral insights will innovatively be applied in the leadership domain. Strategic followership recommendations in the wake of digitalization will be given. The application of behavioral economics to leadership and digital governance accounts for the most cutting-edge approach to capture the power of real-world relevant economics.

The book innovates strategic leadership theory for wise followership insights—stressing the case of a combined decision-making whether one wants to be a leader or follower throughout the different choice architectures presented in a complex world. Drawing from a line of research on bounded rationality, the book will enable readers to empirically find how economics can better societal conditions. The book also provides clear guidelines how to use digital tools wisely with a future-oriented and ethical view. Individual decision makers are also empowered to maximize financial profits thanks to a clear understanding of the impact of heuristics and biases as well as social influences. Behavioral strategists will learn how to lead other people by nudges and winks to choose wisely for themselves, others, and society. Conscientious consumers will be advised how to navigate through a constantly evolving online world. Global governance specialists will learn how information may distort economic rational choice.

The concept of "Conscientious Usership" is introduced—capturing unpaid online usership of search engines, social media, and online market places that gets harvested for big data. The use of future-oriented discounting in the predicament between utility in information sharing versus dignity in privacy in the digital millennium is emphasized. Special attention is paid to digital social media platforms' role as information exchange on minor medical conditions during the COVID-19 pandemic—foremost in social media COVID Long Haul groups that currently revolutionize the medical information market but come with the caveat of privacy concerns and discrimination potential of vulnerable patient groups.

Data ownership but also the culture of e-ethics that we all create online will be emphasized in the digital age.

The book will start by introducing the wide range of psychological, economic, and sociological laboratory and field experiments that proved human beings deviating from rational choices. Standard neoclassical profit maximization axioms were found to fail to explain how humans actually behave. Human beings rather use heuristics in day-to-day decision-making. These mental short cuts enable us to cope with a complex world yet also often leave individuals biased and falling ashtray to decision-making failures. What followed was the powerful extension of these behavioral insights for public administration and public policy-making. Behavioral economists proposed to nudge and wink citizens to make better choices for them and the community. Many different applications of rational coordination followed ranging from improved organ donations, health, wealth, and time management, to name a few. The results of these studies will be presented in order to derive hands-on leadership recommendations but also strategic followership guidance.

The book innovatively captures behavioral economics and proposes further analysis strategies to unravel how to use economics for the greater societal good. The book enables the reader to empirically find how economics can improve societal conditions with particular attention to digitalization. By drawing from the historical foundations of political economy, the book seeks to advance the field of behavioral economics through a critical stance on behavioral sciences' use for guiding on public concerns.

While acknowledging that behavioral economics revolutionized mainstream neoclassical economics, the limitations and implicit problems arising from nudging will be uncovered in academic but also applied and quasi-Socratic writing. Throughout the book, readers are guided to investigate and scientifically propose further analysis strategies to unravel how the use of economics for the greater societal good can be improved. The book thereby takes a heterodox economics stance in order to search for interdisciplinary improvement recommendations for the use of economics in global governance.

The book proposes an international focus on behavioral insights by providing information about the state of the art of behavioral traditions in Europe and North America. Stakeholder-specific facets of behavioral sciences and the different scientific disciplines' approaches toward heterodox economics and behavioral finance are provided in the search for governance recommendations. The role of social online media for financial markets will be outlined with particular attention to social volatility. Delineating the potential of behavioral economics to implement social welfare portrays economics as a real-world relevant means to minimize societal downfalls. Informing about the most novel behavioral insights in the finance domain helps readers to imbue trust in the globalized world economy.

In a nested approach, the book features qualitative and quantitative research projects to gain invaluable information about the interaction of economic market with the real-world economy. The book is targeted to have direct implications for policy-makers. The monograph can also teach upcoming behavioral economists on how to conduct heterodox science projects alongside breeding civic engagement and leadership on behavioral economics and finance leadership.

Ethics of inclusion and economics of diversity in the post-COVID-19 era are covered. Particular focus will be paid to COVID-induced and pandemic-exacerbated inequality drawing from concrete cases on the provision of equal access to health, economic, and finance market benefits as well as digitalization benefits. In light of COVID-19-induced market instability and the potential long COVID threat, the need for integrating rest and recovery in economic calculus is argued for. Artificial intelligence, robotics but also big data insights appear as digital remedies to aid on pandemic prevention but also human recovery assistance in the aftermath of the pandemic. The need for research on healthcare-dependent multipliers within economics is made.

The concept of e-ethics will be expanded on behavioral online ethics. Edges of ethical market behavior in the harmful online abuse of nudges in negative blackhat strategies in search engines will be thematized. Blackhat strategies are used to make competitors' online content disappear and search engine results clogged with useless information or defamation occurs in bloating negative contents via click farms. A fair searchplace mandate and the concept of searchplace discrimination alleviation will be argued for. A "Right to Not be Forgotten" and to be online present is argued for in the eye of online searchplace discrimination.

A future of behavioral economics' perspective covers the most vibrant extensions in behavioral law and economics with particular attention to disparate impact analyses. Concrete contemporary examples are given, such as the current International Law Commission opening up the field of behavioral law in the reality check of law-in-books versus law-in-practice. The second edition of *Behavioral Economics and Finance Leadership: Nudging and Winking to Make Better Choices* includes a revision of references throughout the book and most novel literature in the field.

Overall, the monograph strives to teach the necessary leadership skills to bring change in a world of nudges, winks, and artificial intelligence encroaching human workspaces. Delineating the potential of behavioral economics to implement social welfare portrays economics as a real-world relevant means to minimize societal downfalls and imbue trust in the globalized world economy. Alongside providing an overview of behavioral sciences with an application in leadership and strategic followership, the book will also take a critical approach to the economic analysis of contemporary public choices.

This book developed during placements at Harvard University, Columbia University, Princeton University, Yale University, and the New School. The author acknowledges support of the Fulbright Commission, the Max Kade Foundation New York in cooperation with the Austrian Academy of Sciences, the Prize Fellowship in the Inter-University Consortium of New York, the Schwartz Center for Economic Policy Analysis, George Washington University, the International Institute for Applied Systems Analysis, the Fritz Thyssen Foundation, the University of Vienna and the Vienna University of Economics and Business. At the moment of the second edition, the author was primarily affiliated with Columbia University and conducted an academic training with the Friends of the International Institute for Applied Systems Analysis (IIASA) in cooperation with the United States National Academy of Sciences. The author declares no conflict of interest. All omissions, errors, and misunderstandings in this piece are solely the author's.

New York, USA
July 2022

Julia Puaschunder

Contents

Part I

Behavioral Foundations

Behavioral Economics 1

1.1 Introduction

Since the end of the 1970s, a wide range of psychological, economic, and sociological laboratory and field experiments proved human beings deviating from rational choices. Standard neoclassical profit maximization axioms were outlined to fail to explain how humans actually behave. Human beings were rather found to use heuristics in day-to-day decision-making. These mental shortcuts enable us to cope with information overload in a complex world. While in Europe heuristics were more likely to be seen as evolutionary-grown decision-making aids to navigate a stressful environment, foremost North American behavioral economists proposed to nudge and wink citizens to make better choices with many different applications in multiple domains.

This book will (1) start with a review of the contemporary literature on human decision-making failures in Europe and North America presenting theoretical foundations and a wide range of nudges and winks developed to curb harmful consequences of humane decision-making fallibility. The book will then (2) propose how to use mental heuristics, biases, and nudges in the finance domain to profit from economic markets providing clear communication strategies for financial executives. (3) Behavioral ethics toward inclusion and the behavioral economics of diversity will be presented as most novel advancement in the field of behavioral leadership. The monograph finishes with (4) clear leadership and followership directives on nudging in the digital age with particular attention to wise usership and the most recent developments on social media online data sharing in our post-COVID-19 era.

Part 1: Since the end of the 1970s, behavioral economics revolutionized mainstream neoclassical economics and revolutionized decision-making theory. Behavioral economists have recently started to nudge—and most recently wink—people into favorable decision outcomes, offering promising avenues to steer social

J. Puaschunder, *Advances in Behavioral Economics and Finance Leadership*,
Contributions to Economics, https://doi.org/10.1007/978-3-031-15710-3_1

responsibility in very many different domains, ranging from marketing, corporate governance to public affairs, and most recently leadership. A wide range of psychological, economic, and sociological laboratory and field experiments proved human beings deviating from rational choices as standard neoclassical profit maximization axioms failed to explain how humans actually behave. Human beings rather use heuristics in their day-to-day decision-making. These mental shortcuts enable us to cope with a complex world yet also often leave individuals biased and falling astray to decision-making failures.

What followed was the powerful extension of these behavioral insights in the domains of public administration and public policy-making. Behavioral economists proposed to nudge and wink citizens to make better choices for them and the community. Many different applications of rational and efficient coordination followed ranging from improved organ donations, health, wealth, and time management, to name a few.

The first part of the book will be dedicated to describing what nudges and winks are. A behavioral economics overview will be blended with an introduction of behavioral insights in financial markets and international trade. Part 1 closes with an international outlook comparing the North American with the European school of behavioral economics. Whereas in the United States and Canada, human decision-making heuristics are often seen as a failure and advice is given on how to improve decision-making outcomes toward rationality; the European school portrays these evolutionary-grown decision techniques as humane helpful strategies to cope with an overly complex environment and focus aids on humane features of life. Most recently, arguments have been made that behavioral economists should rather inform people about the likelihood of outcomes and consequences in order to empower them to choose according to their personal preferences in a democratization of behavioral law and economics (Liscow & Markovits 2022). Ethics of inclusion, law and economics interdisciplinary dialogue building, and human-artificial intelligence algorithm compatibility are expected to become key advancements in behavioral economics and finance leadership of the future.

Part 2: Behavioral finance is one of the most novel developments in behavioral economics. Since the end of the 1970s, a wide range of psychological, economic, and sociological laboratory and field experiments proved human beings deviating from rational choices and standard neoclassical profit maximization axioms that fail to explain how humans actually behave. Human beings were rather found to use heuristics in the day-to-day decision-making. These mental shortcuts enable us to cope with information overload in a complex world. Behavioral economists proposed to nudge and wink citizens to make better choices for them with many different applications in very many different domains. Nudges are verbal cues that propose positive reinforcement and indirect suggestions as ways to influence the behavior and decision-making of groups or individuals. Winks are nonverbal cues that try to elicit certain acts and choices of humans.

The book provides international and interdisciplinary perspectives on economics and finance in the age of digital inequality. A multi-dimensional analysis will outline the qualitative and quantitative facets of digital inequality. The size, scope and dimensions of inequality will be explored in a truly multi-modal way covering different disciplinary angles and various regions of the world. Discussing digital inequality from different angles for a broad audience is meant to provide wide-reaching insights for an international audiences, given the global outlook. The international perspective predestines the book to be of value for global governance scholars or corporate executives, who focus on driving cutting-edge innovations. The monograph also offers a positive approach to innovation and features multi-disciplinary analyses. A first analysis of the book will shed light at the advantages of digitalization with particular attention to behavioral economics. The book will then also cover rather unknown digitalization disruption's consequences to widen existing inequality gaps or even open up novel, potentially unforeseen inequalities. Outlining the blatant but also hidden inequalities imbued in digitalization within society between countries but also over time will bring increasing and lasting value to historical research.

In all this literature missing is clear information on how to lead efficiently given mental shortcuts and behavioral biases in a complex world. Completely overlooked has been the strategic followership, hence how to maximize one's profit given we are all leaders and followers alike. But when to become a leader and when to be a strategic follower as well as how to maximize one's decision outcomes as followers are completely unrepresented in the standard basic behavioral economics literature.

Part 2 of the following monograph reviews and proposes how to use mental heuristics, biases, and nudges in the finance domain in self-leadership but also strategic followership to profit from economic markets while paying attention to how heuristics and biases shape economic systems. In addition, the role of social online media for finance and nudging and winking consumer choices via fast-paced online platforms will be thematized with particular attention to novel and unforeseen digital inequalities. The most recent trend of arguing for a democratization of online worktime on social media will be discussed.

Part 3 aims at nurturing interdisciplinary interests in behavioral economics innovatively applied in the leadership domain in the artificial intelligence age. The application of behavioral economics to leadership and governance accounts for the most cutting-edge approach to capture the power of real-world relevant economics. Drawing from a line of research on bounded rationality, this part will enable readers to find how economics can better societal conditions and how they can lead other people by nudges and winks to better decide for themselves, others, and society during our contemporary age of artificial intelligence encroaching markets. Delineating the potential of behavioral economics to implement social welfare portrays economics as a real-world relevant means to minimize societal downfalls and imbue trust in the globalized world economy. Alongside providing an overview of behavioral sciences with an application in leadership during the artificial intelligence revolution, this part will also take a critical approach to the economic analysis of contemporary public choices and teach the necessary leadership skills to bring change in a world of nudges, winks, and artificial intelligence encroaching human workspaces. Contemporary global developments in international law going into a behavioral direction in a reality check of law-in-books versus law-in-practice will be discussed to lead the way on the most pressing international customary law.

Overall, while acknowledging that behavioral economics revolutionized mainstream neoclassical economics, the limitations and implicit problems arising from nudging will be uncovered in Socratic writing. The book innovatively captures behavioral economics and proposes further analysis strategies to unravel how to use economics for the greater societal good. By drawing from the historical foundations of political economy, the book will seek to advance the field of behavioral economics through a critical stance on behavioral sciences' use for guiding on public concerns. The book will also enable readers to empirically find how economics can improve societal conditions. Stakeholder-specific facets of behavioral sciences and the different scientific disciplines' approaches toward heterodox economics will be outlined in the search for governance recommendations. Delineating the potential of behavioral economics to implement social welfare portrays economics as a real-world relevant means to minimize societal downfalls and imbue trust in the globalized world economy. For instance, digitalized healthcare information exchange has nowadays also proved as a means for a democratization of health and disease remedies when considering the role of online media forums to discuss COVID long-haul issues [154]. Ethics of inclusion of the future may advance online access to health care and a fair and protected information exchange of vulnerable patients.

Overall, the book plays an important role in the evaluation of nudging and its influence on the stability of economic markets, financial stability, and societal systems. Depicting nudging during this unprecedented time of economic change and regulatory reform in our post-COVID-19 pandemic era holds invaluable historic opportunities for leaders on how to strengthen society by nudges but also overcome unknown emergent risks within globalized markets in resilient finance backed by behavioral insights. In its entirety, the monograph is targeted at bestowing market actors with key qualifications to lead and to follow strategically in a complex and digitalizing world. The monograph also aims at empowering social online media users to reclaim their privacy in the digital world and become aware of the hidden inequalities in modern online marketplaces. The book will also shed most novel light on hidden search place discrimination and offer creative ways how to imbue ethics in innovative market transitions.

1.2 Theoretical Background

Since the end of the 1970s, a wide range of psychological, economic, and sociological laboratory and field experiments proved human beings deviating from rational choices and standard neoclassical profit maximization axioms to fail to explain how humans actually behave [77]. Human beings were rather found to use heuristics in the day-to-day decision-making [78, 79]. These mental shortcuts enable us to cope with information overload in a complex world [203].

From there on, the emerging field of behavioral insights targeted at using human heuristics and biases to improve decision-making in different domains ranging from health, wealth, and prosperity [203]. Behavioral economists proposed to nudge and wink citizens to make better choices for them with many different applications. Behavioral insights teams have been formed to advise individual governments around the globe—for instance, Australia, Canada, Colombia, Germany, Italy, the United Kingdom, and the United States [215]. But also intergovernmental entities, such as the European Commission, or global governance institutions, such as the World Bank and the International Monetary Fund, have started using nudges and winks to improve society [215]. In 2017, behavioral insights entering the public policy domain have been acknowledged by the Nobel Prize in Economics being granted to Richard Thaler.[1] Nudges were defined as verbal cues that propose positive reinforcement and indirect suggestions as ways to influence the behavior and decision-making of groups or individuals. Most recently, winks were added as nonverbal cues that try to elicit certain acts and choices of humans.

The book (1) starts with a review of the contemporary literature on human decision-making failures and the wide range of nudges and winks developed to curb harmful consequences of humane decision-making fallibility and stabilize economic market systems in strategic communication; then

(2) proposes how to use mental heuristics, biases, and nudges in the finance domain to profit from economic markets; and then

(3) finishes with a future outlook on nudging in the twenty-first century and the future of nudges in the age of the artificial intelligence revolution after the post-COVID-19 pandemic.

1.3 Individual Decision-Making Under Uncertainty

Behavioral economics revolutionized decision-making theory. By studying human decision-making fallibility and its consequences, behavioral economists argue that people make decisions based on rules of thumb heuristics that dominate human choices [58, 59, 80]. Laboratory experiments have captured heuristics as mental shortcuts easing mentally constrained human in a complex world [25, 175, 190]. Mental shortcuts were outlined to simplify decision-making and substitute difficult questions with easily applicable automatic behavioral reactions [75]. An impressive line of research has shown that heuristics lead to predictable and systematic errors [207].

These cognitive mental shortcuts seem to set humans on a path to erroneous choices, as the North American school of behavioral economics argues. Heuristics

[1] https://www.nobelprize.org/nobel_prizes/economic-sciences/laureates/2017/thaler-facts.html.

were examined as a source of downfalls on rational and socially wise choices given future uncertainty. Heuristics cause people to make choices much faster, but ultimately less logically than more careful, long-form decision-making. These decision-making failures were studied in order to improve human decision-making outcomes over time and in groups [23].

1.3.1 Homo Nudgiens

Behavioral economists have recently started to nudge—and most recently wink—people into favorable decision outcomes, offering promising avenues to steer social responsibility in public affairs [5, 75]. Individuals were nudged into doing things they naturally would not have a natural tendency of doing. Most recently, behavioral economics innovatively became applied in the public administration and policy domains as a cutting-edge approach to capture the power of real-world relevant economics for improving society. Drawing from a line of research on bounded rationality, behavioral economics accounts for one of the most prominent approaches to minimize societal downfalls and implement social welfare maximization. Behavioral economics is thereby seen as a real-world relevant means to enable global governance in the world economy and improve societal advancement on a global scale [214]. Yet questionable is whether or not economic calculus can be applied to the governance of human activity within society without philosophical foundation or ethical oversight [126]. Heuristics may be studied to help explain why people may act illogical and how their fast and impulsive decision-making can hurt them on the long run and/or in social contexts.

1.3.2 Bounded Rationality and Ethicality

In an impressive line of experiments and field studies, the growing field of behavioral finance has offered behavioral insights on how markets deviate from efficiency. Human actors are prone to base their investment choices on very many other factors than simply volatility and profit maximization opportunities [138]. The book will review some of the behavior insights gained in the last decades and show ways of how to profit from heuristics and biases.

Most recently, nudging has started using the emerging insights about human heuristics and biases to improve decision-making in different domains ranging from health, wealth, and prosperity [203]. Behavioral insights teams have been formed to advise individual governments—e.g., Australia, Austria, Canada, Colombia, Germany, Italy, the United Kingdom, and the United States, intergovernmental entities—e.g., the European Commission, or global governance institutions, such as the World Bank and the International Monetary Fund [215].

While standard microeconomic theory captures exponential temporal discounting to explain rational decision-making, behavioral economics finds human time perception biased by heuristics, analogical thinking, and minimized effort [8, 21,

23, 32, 46, 56, 75, 76, 120, 155, 174, 176–178, 181, 188, 189, 217]. People's cognitive capacities to consider future outcomes in today's decisions are limited [43, 88, 96, 163, 163].

As an overview of renowned heuristics, anchoring speaks about setting an unconnected anchor as a means to change perception. For instance, if negotiating a salary, the offer may be higher if the employee triggered a higher number at the beginning of the conversation, even if that number being completely unrelated to the discussion.

Availability changes people's perception insofar as the more available information is, the more likely people will perceive this information to happen, even if this overinflates their likelihood estimation. For instance, after September 11 car accidents spiked in the US as after the extensive media coverage of aircrafts crashing, Americans started driving long distance rather than taking flights in fear of a plane crash. Driving cars long distance holds a higher risk of car accidents though then aircraft incidents.

Representativeness heuristic is used when making judgments about the probability of an event under uncertainty. Representativeness occurs if a certain characteristic is likely related to another, which then biases people jumping into the conclusion that these characteristics will always be connected. For instance, Austrians appear to have a brought up a history of classical musicians yet not every Austrian is a good musician or likes classical music, which is often assumed. Joint decision-making by presenting alternatives together and not as singles were shown to alleviate the representativeness bias [153].

Sunk costs are expenses already done leading to future preferences to continue spending on the cause. An unhappy marriage can be seen as a sunk cost trap, in which the parties involved continue hanging in there for the sake of all the costs already spent together in terms of time and mutual giving to each other.

Familiarity occurs if people prefer things that are similar to them. For instance, you like those with the same first letter as in your name more than others. The familiarity bias is dependent on the cognitive load. The more individuals are distracted, the more we follow the same. For example, a befriended couple of the author got to know each other as being the only obvious representatives of US citizens on a remote airport in Asia in a chaos over a canceled flight. They did not know each other but naturally felt to connect to exchange information in a place where there was no Internet and hardly anyone speaking English as the cognitive load in a foreign environment and stress triggered the search for similarity.

Related is the similarity heuristic, which can lead from prototypes to stereotyping. People attribute attitudes based on a prototype and shun out information that may be contrary to their expectation, especially under time constraints and in an overly complex environment.

The simulation heuristic is a simplified mental strategy, according to which people determine the likelihood of an event based on how easy it is to picture the event mentally.

The fluency heuristic describes that the more fluent a memory is, the more solid it sticks to the memory.

Recognition through memory speaks about individuals valuing those objects they recognize higher than objects they do not recognize.

Social norms are learned throughout life and cultivate our actions. Social norms can lead to a comparison with others and negative herd behavior as obvious in financial market overreactions. Social proof appears to be strong in manipulating our behavior, leading the herd behavior, which can be harmful when resulting in negative groupthink.

Framing is a cognitive heuristic in which people tend to reach conclusions based on the "framework" within which a situation was presented. So, for instance, the likelihood of success or failure of a drug remain the same if we either consider the drug to be successful in 20% and unsuccessful in 80% of all cases. However, the likelihood of people signing up for the drug increases if only being told that the drug is in 20% of all cases successful. So positive and negative frames and a particular focus on only one side influences peoples' behavior.

There is an overall dependence on emotions, mood, weather, color, hygiene, sight, status, and trust, and very many different unrelated things that determine outcomes. Contagion occurs if previous contents, heat or color of the environment bleed into decisions and change our behavior by even completely unrelated events and circumstances. Especially affect seems to be influencing human decision-making. Emotions tend to influence our choices drastically.

There is the so-called primacy-recency effect that emphasizes that the first and last impressions are the most memorable in a flow of information. So the beginning and ending of a presentation are the most memorable parts. Studying right before going to bed also emphasizes memory advantages as no more information from active engagement with the world follows once having fallen asleep.

The endowment effect describes that the mere ownership of an object raises the value of an object. If people are given an object for free, they rate the value higher than if they were asked to purchase the object and name a price for obtaining the good.

Scarcity creates the effect of overestimating the value. In targeted advertisement and offers, it is often mentioned that this is the last chance to sign up or an estimate on how many slots are still available is given. Online shopping often releases information on availability in real time or features a ticker how many slots are available and how many other individuals are currently checking the purchase option in order to emphasize scarcity. Scarcity leads to a price spike also in artistic work. For instance, if a contemporary artist dies, it becomes obvious that the artist will never produce any further work. The sudden ceasing of an artist (e.g., rock stars or painters dying young and/or unexpectedly) usually leads to a demand spike and often to an elevated price of art thereafter. The sudden disappearance of art work can also create a similar scarcity effect coupled with media presence that drive up the artwork or all products of a particular artist (e.g., the disappearance of Leonardo da Vinci's Mona Lisa painting in 1911 or the Cellini's salt cellar Saliera in 2003 from the Kunsthistorische Museum in Vienna, Austria drove up the recognition and price of the artwork and the artists' work in general). Media presence of an artist's work can raise the price of the art in general, even for unrelated reasons. The restitution cases of Nazi looted art improved the price of Egon Schiele and Gustav Klimt's artwork. To this day, unregulated appears so-called chandelier bidding in open auctions, in which auctioneers tend to pretend to see someone raise a paddle in the back of the room during a live auction with the intend to create even more demand and thus eventually pushing up the price by unsubstantiated claims.

Effort speaks to the more effort one puts into a task, the more value the outcome has. Think of a grade you have been working for a long time, which becomes more valuable than an unexpected windfall gain.

The gaze heuristic is used in directing correct motion to achieve a goal using a learned cascade of motions. An example of the gaze heuristic is catching a ball. Gaze as fast gut reaction helps humans and animals process large amounts of information quickly and react, regardless of whether the information is consciously processed.

Take-the-best describes that individuals choose the alternative based on the first cue that discriminates between the alternatives. They tend to take what is closely available.

Decisions in market situations have been found to be overconfident and myopic. People reverse preferences over time and are not consistent in following through with their plans. Laibson's [88] hyperbolically decreasing discounting functions describe more accurately choice behavior of individuals, who tend to be impatient for smaller rewards now rather than waiting for larger ones later [e.g., 3, 18, 43, 51, 52, 61, 62, 64, 66, 84, 86, 88, 97, 103, 107, 115, 116, 157, 194]. Dynamically inconsistent preferences reverse as people are patient when deciding for the future and impatient when choosing for now [70, 88, 108, 113, 164, 200]. Field and laboratory experiments provide widespread empirical evidence for hyperbolic discounting and self-control failures [52, 69, 174, 179] on money management [6, 27, 33, 65, 83, 86, 88, 89, 168, 191, 202, 211], financial benefits [22], credit card debt [109, 184], medical adherence [206], public health [20, 24, 29, 44, 71, 209], addiction [13, 18, 68, 90, 101, 124], social security [106], fiscal policies [82], commitment [45, 175, 179], health exercise [39, 40], employment [41], procrastination [165], diet [163], subscription discipline [122], animal care [61, 107], and consumption [163, 212]. Failures to disciplinedly stick to plans for giving into immediate desires [4, 161, 195] are explained by people caring less about future outcomes in the eye of future uncertainty [99, 180], perceived risk [1995], and transaction costs [30, 49, 52, 85, 107, 160]. Presenting temporal snapshots for now and later concurrently helps overcome myopia [153].

Behavioral insights informing others about their decision-making fallibility and aiding the general public to make more rational choices is rooted in social responsibility and behavioral ethics. Dating back to antique and religious roots, responsibility is foremost described in philosophical, legal, and libertarian writings. The need for a sense of responsibility for societal well-being is attributed to ancient Greek philosophers [186]. Kant [81] introduced responsibility into modern philosophy as an internal moral mainspring for ethicality and universal privilege of society. Kant's categorical imperative addresses social responsibility by advising to solely act in ways one wants to be treated by others. Legal writings outline responsibility as a feature of ethicality and an expression of natural rights.

Evolutionary psychologists outline responsibility as an essential feature of human nature that connects the individual with society. Based on the ethics of morality, responsibility depicts an internal care for others that stems from altruism. Altruism implies benefiting others while lowering the level of personal fitness [205]. Underlying motives for altruism are self-fulfillment by contributing to matters beyond the self and the so-called warm glow—a positive emotion attributed to caring for and giving to others [9, 73]. In an unconscious search for meaning and warm glow, humans exhibit responsible behavior.

Smith [192] and Marx [104] are the first to depict the relation of autonomy and self-determination of working situations leading to self-fulfilling responsibility as a central motivating factor [185]. When responsibility is connected to profound reasons that are central to a persons' identity, individuals identify themselves with these purposes and feel obliged to act accordingly. Self-imposed responsibility goals become compelling drivers for actions and can even leverage into professional endeavors [36, 54]. For example, in the case of social entrepreneurship, responsibility lets social entrepreneurs apply economic acumen to societal causes about which they personally care.

Within groups, individuals are implicitly evaluated on their exhibited social responsibility. Pro-social behavior is triggered by positive reinforcement of peers. In society, responsibility leads to an overall beneficial social climate. Collective responsible caring breeds the so-called social glue—an implicit form of societal order beyond regulations and legal enforcement. The social glue is a prerequisite for collective trust, societal progress, and economic stability. The social glue may also be an evolutionary basis for governmental regulations. However, the social glue is a phenomenon that is primarily descriptively outlined; therefore, almost no data exists for the size of groups and other external factors influencing the social glue.

In the corporate domain, social responsibility is a feature of business ethics. Business ethics foster the responsibility of corporate actors by setting moral anchors in corporate codes of conduct. Corporate codes of ethics constitute norms of what is right, just, and fair that reflect the law and moral convictions of society [67]. When business ethics match personal responsibility endeavors with the corporate culture, work becomes a self-actualizing motivation factor that fosters productivity [31, 57, 186]. Exhibiting commonly-shared social responsibility in line with business ethics strengthens group cohesion and the ability to work harmoniously [53].

Most contemporary responsibilities have been argued in the "ethics of inclusion." Corporate social justice approaches require corporations to take a comparative approach to understand the most contemporary responsibility challenges of our time. In our post-pandemic era, fairness and social justice are thereby often demanded from an ethics perspective. Most recent law and economics developments present ethical dilemmas of inequality arising in equal access to health care, fair economics and finance benefits, access to education, digitalization equity and a fair share of environmentally-favorable conditions within society, around the world but also over time.

Together corporate social justice pledges envision a transition to a more inclusive society. The US Green New Deal and European Green Deal including a sustainable finance taxonomy are thereby advocating for a transition to a green economy and innovative redistribution schemes. Ethics of inclusion serve as a spring feather of equality and social justice heralding in our post-pandemic Renaissance.

An extraordinary obligation to responsibility is attributed to leaders as for serving as role models and incorporating the aspirations of multiple constituents, whose wants and needs leaders balance [93, 117]. Corporate, financial, and political leadership disproportionately impacts on the lives of present and future generations [94]. More than others are leaders change agents who can foresee and respond efficiently to future anticipations. Leaders' hierarchical positioning and established power bases grant the opportunity to institutionalize responsibility and bestow subordinates with transformational aspiration of responsibility [10, 19, 38, 94]. Psychologists and anthropologists point out individuals' ability to learn responsibility from role models. Within society, political leaders and institutional role models hold the potential to ignite and strengthen social responsibility. As such, in his historic inauguration speech, the US president Barack Obama openly called for responsibility [55, 125] (Washington Post, January 21, 2009). Concurrently World Bank president Robert Zoellick addressed the "new era of responsibility" featuring sustainable globalization (Financial Times, January 25, 2009). In the face of the 2008 world financial crises, both these role models call for responsibility in their public addresses as an underpinning of any vital and modern market economy. In addition to political leaders, academics and intellectual leaders have shed light on disciplined democracies in the age of global capitalism [26].

Paying attention to responsible leadership is beneficial for leaders and society: When leaders' decisions are subject to public scrutiny, responsibility enhances the constituents' acceptance of decisions, which fosters their implementation and in reverse contributes to the success of political leaders. For corporate leaders, responsibility is a crucial component of business performance as for strengthening employee motivation and ensuring long-term positive stakeholder relations [94]. Financial ethical leadership is an implicit means of reducing the likelihood of fiduciary breaches in principal-agent relations.

As a part of the human nature, responsibility underlies fallibility. Responsibility deficiencies arise when moral individuals are not aware that their decision-making implies ethical considerations or negative societal externalities. If individuals make moral judgments quickly and intuitively—solely based on their gut feelings of right and wrong—they are prone to believe that their behavior is ethical, yet at the same time fall prey to unconscious biases and accidental unethicality [63]. Based on Simon's concept of bounded rationality—comprising a model in which human rationality is very much bounded by the situation and mental capacity limitations—bounded ethicality attributes human fallibility in ethical decisions [102, 114, 187]. Bounded ethicality describes unintentional deviations from morality and ethical behavior, by which systemic and predictable psychological processes lead people to engage in ethically questionable behavior inconsistent with their conscious ethical notion. Recently bounded ethicality has become subject to descriptive and experimental scrutiny. An exploratory examination of bounded ethicality phenomena is provided in Table 1.1.

Exploratory evidence describes bounded ethicality to stem from uncertainty and information deficiencies in the way we collect, process, and use information in our decision-making. Decisions are made behind a "veil of ignorance" of perceivable information (*bounded awareness*) due to mental capacity limitations and biased cognitions (*ethical fading*) [100, p. 91, 159, 207].

We tend to ignore unethicality cues regarding ourselves (*ethical detachment, moral disengagement*). As we are prone to adjusting our self-perception to meet our conscious ethical standards (*memory revisionism*), unethicality can unnoticingly slip into our behavior (*slippery slope*).

With a focus on the present (*availability heuristic*), we ignore future generations as potential victims of today's consumption (*discounting the future, intergenerational anonymity, temporal trade-off predicaments*). Our tendency to reside in the status quo (*status quo bias*) and adapt our perception and needs to slightly change conditions (*tuning*) makes us vulnerable to lurking crises. This is especially crucial in sustainability considerations, in which we tend to underestimate the exponential cost increase of environmental protection and climate change [137]. When refraining from discipline today, we miss opportunities for improving our future conditions (*want/should conflicts*).

Discounting future events includes perspectives about the future into our everyday economic decision-making as a prerequisite for sustainable development. However, if we anticipate the future in discounting, we hold positive illusions (*behavioral forecasting errors, positive illusions*) and *future uncertainty* exacerbates this view.

We have misperceptions and a biased judgment of others' ethicality. When it comes to others, we unreasonably raise the standards of responsibility (*fairness bias*) and selfishly claim disproportionate shares of common goods (*reactive egoism*). In negotiations, we fail to see alternative solutions to ethical predicaments (*mythical fixed pie of negotiation*).

Collectively-shared responsibility gets blurred among constituents (*bounded* and *collective responsibility*)—leaving the individual with inefficient rigor to fulfilling externally imposed, standardized goals (*negative goal attainment*). Citizens' *public goods detachment* and *mispriced values of natural resources* hinder environmental protection. Organizational command structures and information deficiencies impact on subordinates' social responsibility (*indirect agency, principal, agent predicaments*). Too aspirational goals hinder individual's actions to pursue them or act upon related other goals (*aspirational goals fallibility*).

Table 1.1 Overview of bounded ethicality phenomena

Bounded ethicality		
Information deficiencies	• ***Bounded awareness*** refers to systematic information deficiencies that prevent individuals from noticing or focusing on relevant data	• ***Bounded awareness***: As a misalignment between the information needed for a rational decision and the information included in the standard decision processes, bounded awareness refers to systematic information deficiencies that prevent people from noticing or focusing on useful, observable, and relevant data. Bounded awareness is most likely to occur when the unethical behavior is unclear, distant, and erodes over time. The incremental manner, in which the information is neglected, increases the likelihood to fall prey to predictable surprises. Boundedness even holds for beholders who have the tendency to overlook unethicality in others.
	• ***Ethical fading*** is a process by which a person does not realize that decisions have ethical implications	• ***Ethical fading***: The extent to which decision-makers perceive the ethical aspects of decisions affects their behavior. In the case of ethical fading, individuals do not realize that decisions have an ethical impact as moral implications and negative externalities are not activated at the time of the decision-making [198]. Framing and sanctions imposed on unethical behavior moderate the degree to which ethicality is cognitively available or fades.
Ignorance of one's own unethicality	• ***Ethical detachment*** occurs when self-interested individuals have difficulty in approaching ethical dilemmas without self-serving bias	• ***Ethical detachment***: Individuals who hold vested self-interests are prone to approach situations self-biased and tend to ignore their self-centeredness [166]. In conflicts between acting in one's self-interest and being socially responsible, self-serving options appear more natural than ethical considerations, which may include complex meta-cognitions about others and uncertain future perspectives.

(continued)

Table 1.1 (continued)

Bounded ethicality		
	• ***Memory revisionism*** attributes that individuals selectively and egocentrically revise their memory as for threats of perceiving themselves as being unethical	• ***Memory revisionism***: It is a process by which people selectively revise their memory as for threats to the self-image [50, 103]. When values are challenged by actual behavior, individuals' perception gets biases in order to refrain from unethical notions of the self [28]. Through a selective memory filter, individuals search for biased self-supporting evidence [15, 118, 156, 193]. While the selective memory sustains positive self-perceptions, it impedes the ability to strive for higher levels of ethicality [119, 217].
	• ***Moral disengagement*** implies motivated forgetting of moral rules and unethical information cues of the self	• ***Moral disengagement***: Motivated forgetting of moral standards avoids distress of immorality, yet at the same time allows guiltless unethicality to remain uncensored [14, 16]. Moral salience and the need for justification reduce moral disengagement.
	• ***Slippery slope*** refers to a potential decline in ethicality as for a chain of unnoticed deviations from regular moral standards	• ***Slippery slope***: Individuals compare their actions to their past behavior. If their unethicality represents a small deviation from the standard, the change remains unnoticed and a potentially slightly more unethical behavior becomes the new reference point [198]. The potential gradual decline of ethicality follows a slippery slope featuring incremental change. In sync, beholders are more likely to accept others' immorality if their unethical behavior erodes slowly and develops gradually over time rather than occurs in an abrupt shift.

(continued)

Table 1.1 (continued)

Bounded ethicality		
Focus on the present while being blinded by future uncertainty	• The ***Availability heuristic*** lets decision-makers overweight current, vivid happenings	• ***Availability heuristic***: Temporal perspectives impact on decision-making. The distant future is viewed as abstract and subordinate, while the present appears more detailed and concrete. Close events that affect us immediately are more easily recalled and seem more available [207]. The availability heuristic lets decision-makers overweight vivid instances and neglect future considerations. Outcomes that demand the anticipation of the future are less vivid, harder to process, and require more involvement. When outcomes are not vivid, humans tend to avoid action. The lack of vividness of environmental decay and climate change makes decision-makers refrain from noticing future resource constraints and intergenerational fairness predicaments [149]. Falling prey to the availability heuristic is especially crucial for corporate and political leaders, whose decision-making extraordinarily impacts on future societal conditions.
	• ***Intergenerational anonymity*** holds that individuals have difficulty envisioning future generations as potential victims of today's consumption	• ***Intergenerational anonymity*** holds that individuals have difficulty envisioning anonymous future generations as potential victims of today's consumption. Future generations are a vague group of people living in a distant time. The more distant and uncertain future negative impacts are, the less likely individuals consider them. In the environmental domain, the lack of future anticipation causes neglect of environmental resources preservation [133, 134].

(continued)

Table 1.1 (continued)

Bounded ethicality		
	• The ***Status quo bias*** addresses that individuals strive to maintain their current position which results in an overall ignorance of the necessity to change	• ***Status quo bias***: Individuals have an aversion to deviations from their status quo [170, 199]. The status quo bias addresses that individuals strive to maintain their current position as for information awareness deficiencies and ignorance of the necessity to change. The status quo bias becomes stronger when multiple parties place different weights onto various concerns. In the environmental context, the status quo bias is an irrational barrier to preventing predictable surprises. Governmental officials refraining from improving obsolete policies imply long-term societal constraints.
	• ***Temporal trade-off predicaments*** arise when short-term self-interests stand at odds with long-term societal concerns	• ***Temporal trade-off predicaments***: Dilemmas arise when self-interests stand at odds with societal responsibility. In temporal trade-off predicaments, individuals tend to overweight present consumption at the expense of future generations [210]. When immediate losses loom larger than prospective future gains, we tend to fail to act in time and miss opportunities for creating long-term value. As sustainability policies have immediate costs, their immediate implementation may be refrained from although it would result in net societal savings.

(continued)

Table 1.1 (continued)

Bounded ethicality		
	• ***Discounting the future*** implies that people tend to overweigh short-term considerations rather than evaluating options with a long-term perspective	• ***Discounting the future***: In daily decisions, humans have a general tendency to discount future problems. The discount function resembles the shape of a hyperbola, which implies that people overweight short-term considerations rather than evaluating options from a long-term perspective [52, 87]. Future discounting gets more skewed, the more the uncertain and distant future perspectives are [210]. Future discounting results in overweighting the immediate costs of environmental conscientiousness and people refraining from necessary action [1, 210]. As for policy implementations, citizens were found to support environmental policies with initial costs and long-term benefits when the policies were to be implemented in the future rather than immediately.

(continued)

Table 1.1 (continued)

Bounded ethicality		
	• ***Want/should conflicts*** comprise intrapersonal dilemmas between immediate gratification and future benefits	• ***Want/should conflicts***: While game theory sheds light on economic facets of common goods dilemmas, scarce is the knowledge on concomitant emotional facets of social dilemmas. The *want/should* distinction depicts emotional facets of human decision-making predicaments between egoistic and social responsibility. Hot-headed *want* choices are emotional, affective, and impulsive, while emotionally cooler *should* decisions are characterized as rational, cognitive, and thoughtful. The *want* self is related to self-interests and a disregard of ethical considerations, while the *should* self encompasses ethical intentions and moral principles. *Want* needs feature immediate gratification, while *shoulds* address future benefits [2, 3, 182, 213]. Prior to ethical decision-making, individuals believe to act in line with the *should*; in the aftermath of an unethical decision, they distort the perception of our *want* behavior to hold themselves ethical in line with the *should*. Exploratory evidence indicates that people learn about their *want/should* biases with experience and gain the capacity to curb its influence. As for policy implications, individuals have an increased willingness to support *should* policies when they are to be implemented in the distant rather than in the near future.
	• ***Tuning*** refers to an adaptation of preferences based on availability and circumstances	• ***Tuning***: It is the human tendency to mentally mold preferences, tastes, and habits in accordance with the available resources and circumstances. As for reducing the likelihood of necessary changes, tuning implies vulnerability to predictable surprises of future resource constraints.

(continued)

Table 1.1 (continued)

Bounded ethicality		
Positive illusions about the future	• ***Behavioral forecasting errors*** occur when individuals are overly confident in their future performance	• ***Behavioral forecasting errors***: Human forecasts of behavioral consequences are often inaccurate [121, 183]. In general, individuals are overly confident in their predicted future performance—especially when socially desirable and moral choices are involved [42, 48, 183, 208].
	• ***Positive illusions*** imply unrealistic optimism based on the belief that the future will be better and brighter than possible	• ***Positive illusions***: They describe unrealistic optimism based on the belief that the future will be better and brighter than realistically possible [197]. Positive illusions help to self-protect the self-esteem, enhance commitment and allow people to persist difficult tasks. At the same time, positive illusions reduce rationality [197]. In the false belief of control of uncontrollable events, positive illusions foster ignorance for the urgency of change, which counterweights environmental protection [34].
	• ***Future uncertainty*** exacerbates biases as for difficulty in imagining future options	• ***Future uncertainty***: It crucially impacts on decision-making as for exacerbating ethical ignorance and biases. Lacking imagination of future perspectives forces people to mentally separate decision choices—that is to process them one by one—which fortifies human fallibility. Future uncertainty causes unethicality to remain unnoticed during the decision-making and lets individuals favor irrational *want* choices.

(continued)

Table 1.1 (continued)

Bounded ethicality		
Misperception of others' ethicality and related negotiation biases	• The ***Fairness bias*** attributes that we think that we are fairer than others and more critical of others' ethicality	• ***Fairness bias***: The human view of fairness is biased by self-interest. The fairness bias attributes that we think that we are more honest, trustworthy, and fair than others and we are more critical and suspicious of others' ethicality [7, 17, 47, 48, 110]. Humans preferably notice and attend information that supports their self-interests, recall self-supportive evidence more easily and overweight self-serving views in personal judgments [12, 98, 111, 112, 204].
	• ***Reactive egoism*** occurs when group members, who are made aware of their unfair claim of collective goods, become more egoistic	• ***Reactive egoism***: Group members often claim a selfishly biased unfair share of collective resources and believe that they deserve an unfair credit for collaborative efforts [11, 91, 166]. This cognitive bias can be overcome if individuals are made to consider others' perspectives. However, this perspective-taking invokes cynicism and egoistic behavior—a phenomenon called reactive egoism [11, 123]. Perspective-taking also reduces the interest in future collaborations among those who believe that they contributed more than others. Reactive egoism gets more accentuated in competitive contexts. Competing parties hold the honest belief that it is fair for them to bear less responsibility and act accordingly.

(continued)

Table 1.1 (continued)

Bounded ethicality		
	• ***Mythical fixed pie of negotiation*** depicts negotiators' failure to look for solutions that heighten the pool of resources and/or degrees of freedom	• ***Mythical fixed pie of negotiation***: In negotiations, individuals tend to think that they debate over finite resources. Negotiators thereby fail to look for alternatives that enlarge the pool of resources and degrees of freedom. As for sustainability, the search for environmental protection innovations holds long-term advantages and positive externalities that can easily be overshadowed by narrow-minded, short-term foci [126].
Blurred collective responsibility	• ***Bounded responsibility*** addresses the fact that responsibility gets diffused the more constituents are involved	• ***Bounded responsibility***: Responsibility is most blatant in one-on-one relationships and gets diffused the more constituents are involved [14]. Responsibility gets blurred, the more distant victims and outcomes appear and the less identifiable negative externalities are. The different scales of responsibility and the remedies to counterweight the consequences of biases become apparent when studying the nature and enforcement of responsibility. While in nuclear families responsibility is based on mutual trust, within society social responsibility is foremost regulated by contracts and governmental control.
	• ***Collective responsibility*** implies collectively-shared problems featuring a lowered degree of individuals' attention for individual solution-finding	• ***Collective responsibility***: Collectively-shared problems feature a lowered degree of individuals' attention to finding solutions on an individual basis. If collectively-shared problems are "psychologically outsourced," individuals think that others will take care of their solution. The illusion of others addressing challenges crowds out the individual motivation to tackle problems and change respective behavioral patterns as for their solution.

(continued)

Table 1.1 (continued)

Bounded ethicality		
	• ***Mispricing of natural resources*** refers to underestimating the price of environmental preservation	• ***Mispricing of natural resources***: People view the preservation of environmental assets as voluntary contribution rather than strategic advantages. Future intangible preferences and predicted expected utility are hard to imagine, which reduces the willingness-to-pay for the preservation of environmental goods and leads to long-term societal losses and intergenerational equity constraints [138, 139].
	• ***Public goods detachment*** addresses a hunch that collectively-shared common goods go hand-in-hand with lowered degrees of loss aversion	• ***Public goods detachment***: Individuals have an aversion to losses from their endowments [170, 203]. This loss aversion may be higher for individual property of tangible goods than for intangible common goods. Psychological detachment of ownership of goods is yet to be tested for natural resources.
	• ***Negative goal attainment*** outlines that extrinsically-imposed goals may crowd out intrinsic motivation and efficient, flexible solution-finding	• ***Negative goal attainment***: Externally-imposed standards influence people's decision-making. An over-reliance on regulatory goals creates a dominant wish to meet given standard. However, if these standards are obsolete or inefficient, they can crowd out intrinsic motivation for change and hinder flexible solution-finding. Extrinsically motivating goal attainment can thus lower the motivation to preserve natural resources and overshadow efficient environmental protection [173].

(continued)

Table 1.1 (continued)

Bounded ethicality		
	• In ***Principal-agent predicaments*** leaders direct subordinates to contribute and/or bear responsibility for their unethicality	• ***Principal-agent predicaments***: Collectively-shared responsibilities can blur the focus on ethicality. Diffusion of responsibility becomes most crucial when leaders direct subordinates to contribute to and/or bear responsibility for their unethicality [37].
	• ***Indirect agency*** attributes that agents may indirectly be forced to unethicality by the organizational structure	• ***Indirect agency***: It attributes agents' failure to consider potential unethicality of commands. Agents may indirectly be forced to unethicality by external settings while being blindfold to the ethical consequences of their actions [35]. If agents face ethical infringements, they have the choice of command obedience, resistance, or whistleblowing.
	• ***Aspirational goals***	• ***Aspirational goals***: Aspirational goals that are overpromising can crowd out performance toward that goal and can have negative spill-over effects in other domains. Aspirational goals were found to deter people from acting in their favor. Too overpromising aspirational goals may not only crowd out direct action to achieve these goals, the overambition may also lead to lethargy on other related domains [92].

Unnoticed human decision-making fallibility implies responsibility deficiencies for society. As an avenue for change, "libertarian paternalism" currently promises efficient public policy strategies to surmounting bounded ethicality [203]. Libertarian paternalism features policies that account for how people make decisions and invisibly steer decision-making in a more socially responsible direction. "Paternalism" thereby stands for manipulating people to act according to the preferences of the policy-makers; "libertarian" attributes that changes do not limit individual freedom. By offering "behavioral nudges," policies influence behaviors within the given institutional framework while upholding personal freedom. Examples for nudges are opt-in and opt-out solutions, in which citizens face either actively or passively framed choices that implicitly influence decision outcomes [203]. This implicit guidance has proved as a successful means to trigger organ donations and financial social responsibility [203]. Bounded ethicality is most crucial when social responsibility deficiencies have economic impacts within market systems. Most recent trends demand for a democratization of behavioral choices in behavioral economists only providing information but leaving the decision maker self-empowered choices to be made better informed but on their own [95].

1.3.3 Mental Temporal Accounting

This chapter introduces mental temporal accounting—the behavioral economics application of mental accounting in the time domain. While most discounting studies are in the finance domain, social and environmental components have not gotten much attention.

Time tacts life. While all humans face the same natural constraints of 24-hour days, behavioral economics found individuals differing in discounting preference for immediate rewards over delayed gratification [51, 74, 167, 169]. Regarding monetary gains, individuals were also shown to hold mental accounts dependent on a reference point but also in regard to how to allocate money to causes individuals care about.

But what if individuals also differ in mental temporal accounts, hence regarding how to spend their time? Decision-makers may have natural mental temporal accounts for how to spend 24 h a day, 720 h a month, 8760 h a year, or 700,800 h an average life? Could it be that individuals have implicit mental accounts for how much time to spend on their own, how much time to be allocated toward working, and how much time to just enjoy life in the open environment? If so, could individuals be susceptible to external cues that nudge them into certain mental timeframes that determine their mental time allocation preferences? Could this mental accounting also depend on reference points, such as the age of the individual, and critical life events, such as becoming parents?

If individuals differ in time spent on their own and social time but also vary in their choices of time spent working and time in the natural environment, we could show that the classical mental accounting theory [201] actually describes similar processes as mental temporal accounting how to spend time but also make these insights dependent on the reference points of age and parenthood.

Theoretically, insights about mental temporal accounting may help derive conclusions for contact theory about how social and environmental cues may nudge people to perceive time differently and act on it accordingly. As mental accounting was successfully introduced to be extendable onto time, traditional mental accounting theory [201] should be revisited for attention to time discounting in the social and environmental spheres alongside the economic attention. Elucidating how contexts and experiencing critical life stages of parenthood influence temporal activity allocation choices promises to improve manifold decisions on education, health, asset management, career paths, and common goods preservation throughout life for this generation and the following [150].

Empirical studies tested if the behavioral economics idea of mental accounting in the finance domain and hyperbolic discounting deviations from standard neoclassical discounting functions can be extended to realistically describe how individuals decide in regard to the scarce resource time, which is introduced as mental temporal discounting.

Research question 1 will investigate the existence of certain mental temporal accounting categories, for instance, such as economic, social, and environmental time accounts. If so, research question 2 will capture a discounting variance based on economic, social, and environmental contexts. Research question 3 will test whether there are age-dependent mental temporal accounts and if critical life events of parenthood change time allocations as well as propose *Hypothesis 1* that *the elder one gets, the more future-oriented and pro-social choices become*, which will be introduced as the age paradox in the discussion.

Study 1: *Exploratory meta-analysis of American Time-Use Survey*

A qualitative exploratory meta-analysis content-analyzed time use as mentioned on the American Time Use Survey of the Bureau of Labor Statistics homepage.[2] The American Time Use Survey of the Bureau of Labor Statistics classifies life activities into the following categories: (1) personal care, including sleep; (2) eating and drinking; (3) household activities; (4) purchasing goods and services; (5) caring for and helping household members; (6) caring for and helping non-household members; (7) working and work-related activities; (8) educational activities; (9) organizational, civic, and religious activities; (10) leisure and sports; (11) telephone calls, mail, and e-mail; and (12) other activities, not elsewhere classified.

While these categories grant insights into the general time-use categories, it remains unclear whether some of the categories are considered as social, economic, or environmental activities. For instance, sports can be a social category but also

[2] https://www.bls.gov/tus/.

be performed on one's own. Also, time bracket variant information is not retrieved. For instance, it is not clear whether individuals change their time allocations when considering different time spans.

The lack of information whether the activities are done by oneself or with others demands an empirical investigation of time use to be categorized into social, economic, and environmental times. Study 2 therefore targets at retrieving information on how much time individuals estimate to spend on social, economic, and environmental tasks.

Study 2: *Quantifying time use in the domains of social, economic, and environmental times*

Design: Since the American Time Use Survey of the Bureau of Labor Statistics does not indicate the time categories social, economic, and environmental times, an exploratory study gave quantitative indications of the time use in those specific domains. Social time is defined as time spent with other people and engaging in social interaction, communication, or activities with others. Economic time is defined as time spent using one's labor power and productive capacity, likely to earn money and be or prospectively be a productive part of the labor force. Environmental time is defined as time spent outdoors in the open environment. Overlaps may exist, for instance, if spending time in the environment with others or working outside. Then, respondents are advised to indicate both categories as overlapping times. So, for instance, a person could indicate to spend 10% of a day for economic, 65% of a day for social, and 45% of a day for environmental time, which adds up to 120% time use as for overlapping categories.

Study 2 was conducted over the Internet using the website Amazon Mechanical Turk (MTurk). MTurk is an online labor market in which employers can advertise jobs (typically taking less than 10 min and paying less than $1), and employees can accept posted jobs that are attractive to them. Workers are incentivized based on their performance, which makes MTurk an attractive tool for conducting experiments and surveys [153]. Online labor markets allow us to conduct behavioral experiments in the international arena. Online labor markets use the Internet to connect researchers with potential subjects drawn from an international sample [158]. The entire process of meeting subjects, the survey, and experiment as well as payment of subjects by the researcher occurs over the Internet and computer simulations. The experience is quite similar to performing a set of computer simulations. In literature overviews and replication studies, MTurk has been proven to provide valid and reliable results [72, 158, 196]. With the advent of online markets, many laboratory experiment barriers, such as limited samples and location biases, have been removed. MTurk workers draw samples from all over the world, with the majority of workers being either in the United States or India [158]. The international sample character with an US and Asian sample population makes MTurk an interesting resource pool for conducting experiments on the global validity of time-use decision-making.

Sample: In total, 110 individuals (female = 32 [29.09%], male = 77 [70%], M_{age} = 31, SD_{age} = 8.55, Range = [18, 64]) from around the world participated in the study online. After an informed consent, the online questionnaire investigated time use of men and women. Of the entire sample, 54 individuals (49.1%) indicated to have children and 56 (50.9%) reported that they do not have children. Of the sample with children, 33 respondents had one child (61.11% of the parents' sample, 30% of the total sample), 17 respondents had two children (31.48% of the parents' sample, 16% of the total sample), 3 subjects had 3 children (5.56% of the parents' sample, 2.7% of the total sample), and only 1 survey taker had 4 children (1.85% of the parents' sample, 0.9% of the total sample).

General time-use questions were presented to subjects on Amazon Mechanical Turk in a Qualtrics questionnaire solution. After consenting to a standard informed consent form, all subjects were asked to make an assumption on how much time they spend on: (1) Social time defined as time spent with other people and engaging in social interaction, communication, or activities with others. (2) Economic time defined as time spent using one's labor power and productive capacity, likely to earn money and be or prospectively be a productive part of the labor force. (3) Environmental time defined as time spent outdoors in the open environment. The specific question read: "Please make an assumption how much time you spend for: (1) Social time defined as time spent with other people and engaging in social interaction, communication or activities with others. (2) Economic time defined as time spent using one's labor power and productive capacity, likely to earn money and be or prospectively be a productive part of the labor force. (3) Environmental time defined as time spent outdoors in the open environment on average over a day. The scale below indicates percentages of a day."

The time-use categories were scrambled. All subjects had questions about all-time horizons of a day, week, month, year, and a decade. Time frames and time-use categories' display was scrambled between subjects. Study 2 thereby quantitatively depicted the percentage of time-use estimates between categories per day, week, month, year, and over a decade. The influence of age-varying time-use differences as well as the critical life event of parenthood was studied.

Results: Over all subjects, time is reported to be used differently for social, economic, and environmental times use over a day,[3] a week,[4] a month,[5] a year,[6] and a decade.[7]

---

[3] $t_{S(107)} = 17.050$, df = 106, $p < 0.000$; $t_{Ec(108)} = 28.832$, df = 107, $p < 0.000$; $t_{En(107)} = 39.271$, df = 106, $p < 0.000$.

[4] $t_{S(106)} = 19.011$, df = 105, $p < 0.000$; $t_{Ec(106)} = 29.025$, df = 105, $p < 0.000$; $t_{En(106)} = 14.019$, df = 105, $p < 0.000$.

[5] $t_{S(104)} = 19.219$, df = 103, $p < 0.000$; $t_{Ec(105)} = 27.927$, df = 104, $p < 0.000$; $t_{En(105)} = 14.404$, df = 104, $p < 0.000$.

[6] $t_{S(104)} = 19.752$, df = 103, $p < 0.000$; $t_{Ec(104)} = 28.339$, df = 103, $p < 0.000$; $t_{En(104)} = 14.443$, df = 103, $p < 0.000$.

[7] $t_{S(108)} = 20.600$, df = 107, $p < 0.000$; $t_{Ec(108)} = 28.455$, df = 107, $p < 0.000$; $t_{En(107)} = 15.937$, df = 106, $p < 0.000$.

Over different time horizons, all subjects report different social, economic, and environmental time use. Social time-use perception differs over a day,[8] week,[9] month,[10] year,[11] and a decade.[12] Economic time-use perception differs over a day,[13] week,[14] month,[15] year,[16] and a decade.[17] Environmental time-use perception differs over a day,[18] week,[19] month,[20] year,[21] and a decade.[22] All mean distributions for social, economic, and environmental time-use perception over day, week, month, year, and decade are displayed in Graph 1.1.

Social time-use perception differs over a day,[23] week,[24] month,[25] year,[26] and decade.[27] Economic time-use perception differs over a day,[28] week,[29] month,[30] year,[31] and decade.[32] Environmental time-use perception differs over a day,[33] week,[34] month,[35] year,[36] and decade.[37]

While there are no gender differences to report, age groups make a difference when it comes to time allocation perceptions during a day in the social sphere[38] and the environmental domain,[39] during a week in the social sphere[40] and the

[8] $t_{S(107)} = 17.050$, df $= 106$, $p < 0.000$.
[9] $t_{S(106)} = 19.011$, df $= 105$, $p < 0.000$.
[10] $t_{S(104)} = 19.219$, df $= 103$, $p < 0.000$.
[11] $t_{S(104)} = 19.752$, df $= 103$, $p < 0.000$.
[12] $t_{S(108)} = 20.600$, df $= 107$, $p < 0.000$.
[13] $t_{Ec(108)} = 28.832$, df $= 107$, $p < 0.000$.
[14] $t_{Ec(106)} = 29.025$, df $= 105$, $p < 0.000$.
[15] $t_{Ec(105)} = 27.927$, df $= 104$, $p < 0.000$.
[16] $t_{Ec(104)} = 28.339$, df $= 103$, $p < 0.000$.
[17] $t_{Ec(108)} = 28.455$, df $= 107$, $p < 0.000$.
[18] $t_{En(107)} = 12.979$, df $= 106$, $p < 0.000$.
[19] $t_{En(106)} = 14.019$, df $= 105$, $p < 0.000$.
[20] $t_{En(105)} = 14.404$, df $= 104$, $p < 0.000$.
[21] $t_{En(104)} = 14.443$, df $= 103$, $p < 0.000$.
[22] $t_{En(107)} = 15.937$, df $= 106$, $p < 0.000$.
[23] $\text{Mean}_{(107)} = 40.99$; SD $= 24.890$.
[24] $\text{Mean}_{(106)} = 45.76$; SD $= 24.784$.
[25] $\text{Mean}_{(104)} = 46.13$; SD $= 24.480$.
[26] $\text{Mean}_{(104)} = 46.54$; SD $= 24.028$.
[27] $\text{Mean}_{(104)} = 46.54$; SD $= 24.028$.
[28] $\text{Mean}_{(108)} = 57.99$; SD $= 20.902$.
[29] $\text{Mean}_{(106)} = 56.45$; SD $= 20.025$.
[30] $\text{Mean}_{(105)} = 58.26$; SD $= 21.376$.
[31] $\text{Mean}_{(104)} = 59.18$; SD $= 21.297$.
[32] $\text{Mean}_{(108)} = 56.40$; SD $= 20.598$.
[33] $\text{Mean}_{(107)} = 39.27$; SD $= 31.298$.
[34] $\text{Mean}_{(106)} = 38.91$; SD $= 28.572$.
[35] $\text{Mean}_{(105)} = 40.83$; SD $= 29.046$.
[36] $\text{Mean}_{(104)} = 40.98$; SD $= 28.936$.
[37] $\text{Mean}_{(107)} = 43.34$; SD $= 28.128$.
[38] $F(4, 102) = 2.483$, $p = 0.048$.
[39] $F(4, 102) = 2.414$, $p = 0.054$.
[40] $F(4, 101) = 3.142$, $p = 0.018$.

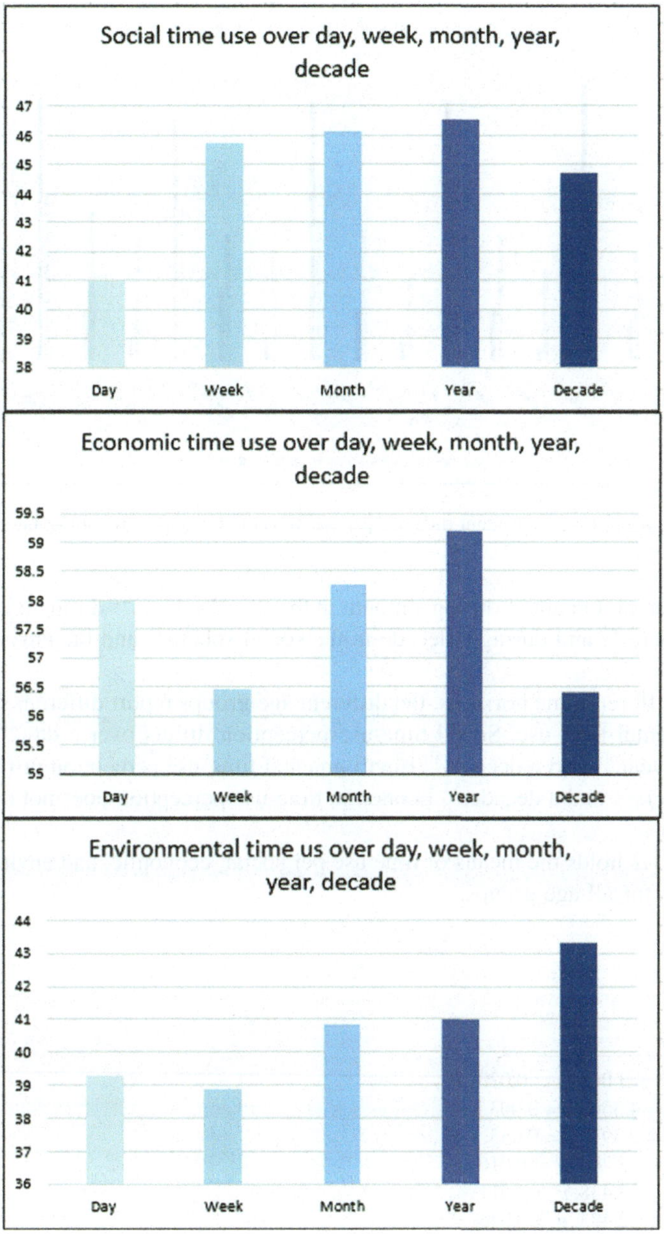

Graph 1.1 Mean distributions for social, economic, and environmental time-use perception over day, week, month, year, and decade

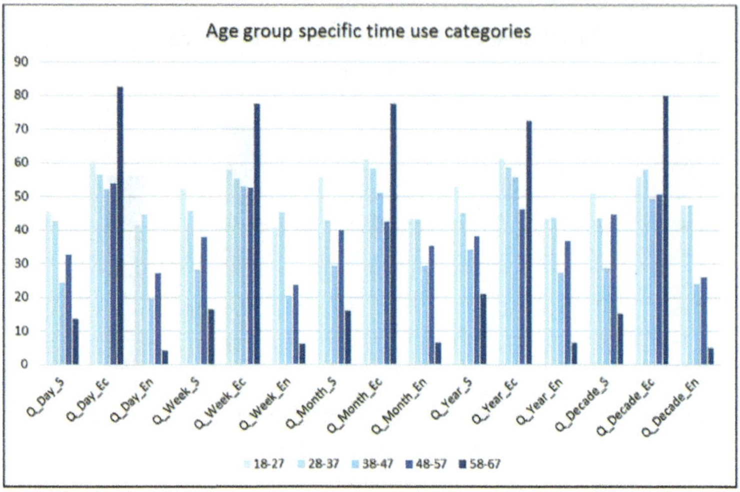

Graph 1.2 Age-group-dependent time use per social, economic, and environmental categories

environmental domain,[41] during a month in the social sphere,[42] during a year in the social sphere,[43] and during a decade in the social sphere[44] and the environmental domain.[45]

Over different time horizons, the different age groups report different social and environmental time use. Social time-use perception differs over a day,[46] week,[47] month,[48] year,[49] and a decade.[50] Environmental time-use perception differs over a day,[51] week,[52] and a decade.[53] Economic time-use perception does not differ over age groups.

Graph 1.2 holds the means of time use per social, economic, and environmental categories for all age groups.

[41] $F(4, 101) = 3.065, p = 0.020$.
[42] $F(4, 99) = 4.356, p = 0.003$.
[43] $F(4, 99) = 2.397, p = 0.055$.
[44] $F(4, 103) = 3.203, p = 0.016$.
[45] $F(4, 102) = 3.438, p = 0.011$.
[46] $F(4, 102) = 2.483, p = 0.048$.
[47] $F(4, 101) = 3.142, p = 0.018$.
[48] $F(4, 99) = 4.356, p = 0.003$.
[49] $F(4, 99) = 2.397, p = 0.055$.
[50] $F(4, 103) = 3.203, p = 0.016$.
[51] $F(4, 102) = 2.414, p = 0.054$.
[52] $F(4, 101) = 3.065, p = 0.020$.
[53] $F(4, 102) = 3.438, p = 0.011$.

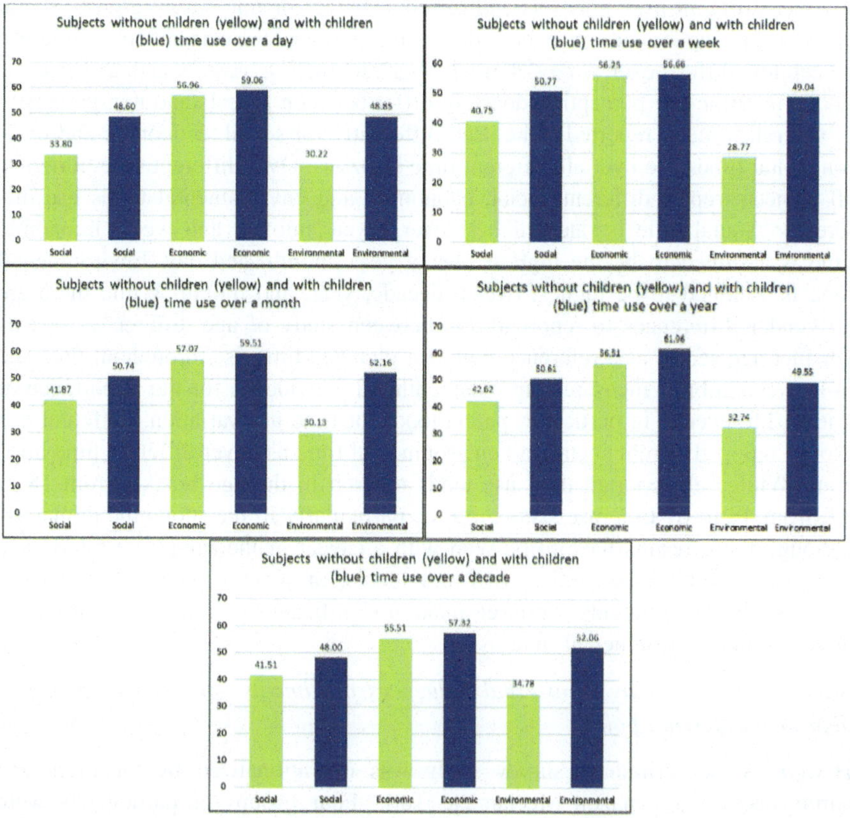

Graph 1.3 Groups of parents or non-parents' time use per social, economic, and environmental categories

Children make a difference when it comes to social[54] and environmental times use during a day;[55] during a week in the social sphere[56] and the environmental domain,[57] during a month in the environmental sphere,[58] during a year in the environmental sphere,[59] and during a decade in the environmental domain.[60] Graph 1.3 holds the means of time use per social, economic, and environmental categories for the group of individuals with children and without children.

---

[54] $t_{S(106)} = -3.208$, df = 105, $p < 0.002$.
[55] $t_{En(106)} = -3.209$, df = 105, $p < 0.002$.
[56] $t_{S(105)} = -2.115$, df = 104, $p < 0.037$.
[57] $t_{En(105)} = -3.889$, df = 104, $p < 0.000$.
[58] $t_{En(104)} = -4.180$, df = 103, $p < 0.000$.
[59] $t_{En(103)} = -3.082$, df = 102, $p < 0.003$.
[60] $t_{En(106)} = -3.324$, df = 105, $p < 0.001$.

Over different time horizons and different parent or non-parent groups, social time-use perception differs over a day[61] and a week.[62] Environmental time-use perception differs over a day,[63] week,[64] a month,[65] a year,[66] and a decade.[67] Economic time-use perception does not differ between parents and non-parents.

Overall, time is reported to be used differently for social, economic, and environmental times use over all different time horizons. Over different time horizons, all subjects report different social, economic, and environmental times use differently. Social time is rated highest over a year, month, and week. Economic time is assumed to be the highest over a year, month, and day. Environmental time is assumed to be highest over a decade, year, and month. While there are no gender differences to report, the in-between study of age differences reveals whether age serves as a reference point for changed time use throughout life. The in-between subject measurement of parenthood is evidence for the importance of critical life events, in particular, parenthood, for time-use variation. Different age groups report different social and environmental time use over different time horizons. While self-reported time use tends to drop in the age bracket from 18 to 47, from 48 to 67 time use seems to rise, particularly in the economic domain—although these results have to be seen with a caveat of the sample of 58–67 age group only being comprised of two individuals. In general, parents report more time use than non-parents. Children make a significant difference when it comes to social and environmental time use.

Study 3: *Survey study of external influences on time-use preferences and age-dependent reference points.*

Design: An experimental survey study was operationalized by Qualtrics and administered via Amazon Mechanical Turk. Four groups of participants were exposed to either (1) social cues (Test group 1), (2) economic cues (Test group 2), (3) environmental cues (Test group 3), or (4) no cues (Control group).

Participants were recruited via Amazon Mechanical Turk. After having been exposed to an informed consent disclaimer, the respondents who agreed to participate were asked to answer an open-ended free-association writing task on either one of the following three conditions in a between-subject design: (1) "Describe your friends (social cue)" (Test condition 1 for Test group 1), (2) "describe your paid work (economic cue)" (Test condition 2 for Test group 2), or (3) "describe a place in nature (environmental cue)" (Test condition 3 for Test group 3). The questions elicited a writing task, in which the respondents were meant to write down free associations after exposure to the cues. The writing task was meant to

[61] $t_{S(106)} = -3.208$, df $= 105$, $p < 0.002$.
[62] $t_{S(105)} = -2.2115$, df $= 104$, $p < 0.037$.
[63] $t_{En(106)} = -3.209$, df $= 105$, $p < 0.002$.
[64] $t_{En(105)} = -3.889$, df $= 104$, $p < 0.000$.
[65] $t_{En(104)} = -4.180$, df $= 103$, $p < 0.000$.
[66] $t_{En(103)} = -3.082$, df $= 102$, $p < 0.003$.
[67] $t_{En(106)} = -3.324$, df $= 105$, $p < 0.001$.

prime respondents into a *social* (Test group 1), *economic* (Test group 2), *environmental* (Test group 3), or *neutral* (Control group) condition. The subjects were split evenly among the three test conditions (between-subject design). The sample from study 2 served as neutral control condition, which did not have any priming of any cues and writing task questions.

Subsequently, all respondents were asked to make an assumption of how much time they spend as: "(1) Social time defined as time spent with other people and engaging in social interaction, communication or activities with others. (2) Economic time defined as time spent using one's labor power and productive capacity, likely to earn money and be or prospectively be a productive part of the labor force. (3) Environmental time defined as time spent outdoors in the open environment." All subjects then had to provide an estimate of how much time in relation to each other they spend on the outlined categories social, economic, and environmental times. Social, economic, and environmental time use of an average over an entire day, an entire week, an entire month, an entire year, and in the last decade of one's life were asked for being estimated.

The questionnaire measured the influence of cues on time allocation preferences and perceptions. The cues served to put the respondents into different social, economic, and environmental mindsets. These external frames were tested for influencing the respondents' time allocation perceptions.

In order to study the impact of situational cues on time allocation choices, the different groups were compared in their general time-use description after having been exposed to a writing task cue. The differences in the time-use description were generated by quantitative responses to the question to estimate their time use per category social, economic, and environmental times. A quantification of contents in relation to each other was pursued in order to derive information on the relative percentage of time-use categories to each other.

After a writing task, general time-use questions were presented to subjects on Amazon Mechanical Turk in a Qualtrics questionnaire solution. After consenting to a standard informed consent form, all subjects were asked to make an assumption on how much time they spend on (1) social time defined as time spent with other people and engaging in social interaction, communication, or activities with others; (2) economic time defined as time spent using one's labor power and productive capacity, likely to earn money and be or prospectively be a productive part of the labor force; and (3) environmental time defined as time spent outdoors in the open environment. The specific question read: "Please make an assumption how much time you spend for: (1) Social time defined as time spent with other people and engaging in social interaction, communication or activities with others. (2) Economic time defined as time spent using one's labor power and productive capacity, likely to earn money and be or prospectively be a productive part of the labor force. (3) Environmental time defined as time spent outdoors in the open environment on average over a day. The scale below indicates percentages of a day."

The time-use categories were scrambled. All subjects had questions about all-time horizons of a day, week, month, year, and a decade. Time frames and time-use categories' display was scrambled between subjects. Study 2 thereby quantitatively depicted the percentage of time-use estimates between categories per day, week, month, year, and over a decade. The influence of age-varying time-use differences as well as the critical life event of parenthood was studied.

The mindframes of social, economic, and environmental factors allowed to draw inferences whether social, economic, or environmental contexts are prone to elicit certain discounting anomalies. The prospective results elucidate whether social, economic, or environmental cues can manipulate time-use preferences. The category priming followed by a free-association writing task also revealed if a certain mindframe either enhances or depletes continuous use of time for the same category.

In addition, the survey asked for the age of the respondent and if they have children and if so, how many children. The time use of different age groups and groups with or without children were compared in order to test for discounting differences and pro-social behavior variations during different time periods throughout life. Children's influence on time allocation preferences as critical life event in parents' lives was examined.

Sample: In total, 262 respondents were included in Study 3. The subjects (female = 75 [28.07%], male = 186 [71.3%], $M_{age} = 30$, $SD_{age} = 8.21$, Range = [18, 65]) from around the world participated in the study online. After an informed consent, the online questionnaire investigated time use of women and men. Of the entire sample, 130 individuals (49.6%) indicated to have children and 132 (50.4%) reported that they do not have children. Of the sample with children, 47 respondents had one child (63.15% of the parents' sample, 17.9% of the total sample), 46 respondents had two children (35.38% of the parents' sample, 17.6% of the total sample), 16 subjects had three children (12.31% of the parents' sample, 6.1% of the total sample), five survey takers had four children (3.85% of the parents' sample, 1.9% of the total sample), and 16 survey takers had five children (12.31% of the parents' sample, 6.1% of the total sample).

Results: Over all subjects and condition and control groups, time is reported to be used differently for social, economic, and environmental times use over a day,[68] a week,[69] a month,[70] a year,[71] and a decade.[72]

[68] $F_S(3, 256) = 9.238$, $p = 0.000$; $F_{EC}(3, 256) = 8.362$, $p = 0.000$; $F_{EN}(3, 256) = 9.442$, $p = 0.000$.

[69] $F_S(3, 254) = 7.421$, $p = 0.000$; $F_{EC}(3, 254) = 10.423$, $p = 0.000$; $F_{EN}(3, 254) = 10.433$, $p = 0.000$.

[70] $F_S(3, 253) = 6.725$, $p = 0.000$; $F_{EC}(3, 253) = 6.402$, $p = 0.000$; $F_{EN}(3, 253) = 6.247$, $p = 0.000$.

[71] $F_S(3, 252) = 7.832$, $p = 0.000$; $F_{EC}(3, 252) = 8.397$, $p = 0.000$; $F_{EN}(3, 252) = 6.810$, $p = 0.000$.

[72] $F_S(3, 256) = 8.576$, $p = 0.000$; $F_{EC}(3, 256) = 8.472$, $p = 0.000$; $F_{EN}(3, 256) = 7.655$, $p = 0.000$.

Discussion: Behavioral economics found individuals to hold mental accounts dependent on a reference point but also in regard to how to allocate money to cause individuals to care about. This chapter presents that individuals also differ in mental temporal accounts, hence regarding how to spend their time. Over all subjects, we find people perceiving time use differently in different time brackets. Individuals tend to have compartments, in which they discount and allocate social, economic, and environmental time use differently. Decision-makers have natural mental temporal accounts for how to spend social, economic, and environmental parts of their lives throughout a day, week, month, year, or decade of an average life. We have implicit mental accounts for how much time to spend on their own, how much time to be allocated toward working, and how much time to just enjoy in the open environment. We can be nudged into different time-use perceptions by external cues. But also the critical life event of parenthood sets us into a different path of spending time in the social, economic, and environmental spheres. It is not a trade-off between categories but rather granting people mental time or more efficient use of time when they become parents.

As a limitation and future research prospect, the found differences of social, economic, and environmental cues impacting on temporal discounting but not public policy choices in the social, economic, and environment sphere demand future investigations of the relation of mental temporal discounting and financial allocation preferences. Mental accounting theory may not easily be extendable on the untested domain of time insofar as the mental cues manipulated time perception but not monetary allocation preferences to the domains of social, economic, and environmental causes. So while we have context-dependent temporal accounting strategies for time, this may not hold for monetary allocation preferences. Time is not money.

The found age dependency of time-use categories demands additional attention. A clear limitation of the study is the narrow sample with a relatively young population. Future research may focus on testing concrete age differences in the time-use preferences with a focus on a more harmoniouslystratified sampling. Specific age categories hold invaluable insights on age groups' specific use of time, which offer precious market implications for very many different industries ranging from consumption goods to service industry, health care, and insurance industries that could serve different age groups more efficiently.

Lastly, the results of environmental cues bestowing with a higher perception of time use in all other domains as well serve as a beautiful case for environmental recreation preservation. Sustainability may grant us a long-term view but meaning of using our time more efficiently in all domains, the environmental but also social and economic spheres as well. In this regard, the results have ample applications, ranging from improving individual's day-to-day decision-making up to intergenerational leadership in light of climate justice demands [129–131, 133–136, 140–148, 151, 153]. The findings hold invaluable insights for improving future-oriented and socially responsible decision-making nudges [127, 128, 132, 134, 144]. The research therefore holds direct applicability for improving the lives of current and future generations in private and public domains [133, 138, 140, 141, 145, 148, 151, 152].

Overall, pointing at the necessity to include past individual points in life into neoclassic economics and hyperbolic discounting spearheads heterodox economics by opening a detected research gap of forward-looking discounting paradigms [153]. Outlining fundamental differences of temporal discounting throughout different ages allows retrieving multi-faceted decision-making influences as well as generating wide-ranging nudges to improve choices over the entire life cycle [203]. The prospective findings also promise to add to contemporary contract theory, which is primarily focused on monetary incentives attention for non-monetary gratification nudges that may imbue motivation to act beyond financial gain-driven ones. Contrasting orthodox temporal discounting with heterodox multi-faceted decision-making approaches that elucidate more exactly how individuals choose to spend time in the course of their lives but also shedding light on the importance of integrating backward-looking aspects in discounting sets the stage for improving future social care beyond one's own existence in a real-world relevant way granting opportunities to imbue eternal equity in humankind.

1.3.4 Evolutionary-Grown Human Decision-Making

The emerging science of heuristics outlines that human decision-making is not efficient but rather subject to unconscious biases. However, the European school on heuristics argues that is it efficient to not process all information provided. Evolutionary arguments propose that all nature is not meeting rational decision-making models. Heuristics are positively portrayed as one mode of thinking that one can actively choose based on the situation and the level of complexity a task demands [60]. Individuals, families, groups, and organizations are shown to often rely on simply heuristics in an adaptive way and willingly ignoring most parts of information available. The task is thereby not to avoid biases but becomes how to weight and adapt to the environment through strategic use of heuristics. Clear strategies are demanded to be put forward how individuals can improve their well-being, wealth, and overall success when understanding heuristics and voluntarily choosing whether or not to apply heuristic shortcuts or more elaborate decision-making. Gigerenzer divides decisions into quantifiable—in accordance with Savage's [171, 172] small world—in which heuristics are better avoided. If decisions are more of a qualitative nature—in accordance with Savage's [171, 172] large world—heuristics are better than optimization and rational maximization strategies.

The given insights will be coupled with the blending of artificial intelligence (AI) aids, providing the first introduction of the role of nudging in the digital age. The big future challenge is to develop a systematic theory of how to use heuristics strategically as well as when and how environmental capacities of the digital age should be added.

References

1. Ackerman, F., & Heinzerling, L. (2004). *Priceless*. New York Press.
2. Ainslie, G. (1975). Specious reward: A behavioral theory of impulsiveness and impulse control. *Psychological Bulletin, 82*(4), 463–496. https://doi.org/10.1037/h0076860
3. Ainslie, G. (1992). *Picoeconomics: The interaction of successive motivational states within the person*. Cambridge University Press.
4. Ainslie, G., & Haslam, N. (1992). Hyperbolic discounting. In G. Loewenstein & I. Elster (Eds.), *Choice over time* (pp. 57–92). Russel Sage.
5. Akerlof, G. A. (2009). *Animal spirits: How human psychology drives the economy, and why it matters for global capitalism*. Princeton University Press.
6. Alberini, A., & Chiabai, A. (2007). Discount rates in risk versus money and money versus money tradeoffs. *Risk Analysis, 27*(2), 483–498. https://doi.org/10.1111/j.1539-6924.2007.00899.x
7. Alicke, M. D. (1985). Global self-evaluation as determined by the desirability and controllability of trait adjectives. *Journal of Personality and Social Psychology, 49*(6), 1621–1630. https://doi.org/10.1037/0022-3514.49.6.1621
8. Allport, G. W. (1979). *The nature of prejudice*. Addison-Wesley.
9. Andreoni, J. (1989). Giving with impure altruism: Applications to charity and Ricardian equivalence. *Journal of Political Economy, 97*, 1447–1458. https://doi.org/10.1086/261662
10. Aronson, E. (2001). Integrating leadership styles and ethical perspectives. *Canadian Journal of Administrative Sciences, 18*(4), 244–256. https://doi.org/10.1111/j.1936-4490.2001.tb00260.x
11. Babcock, L., & Loewenstein, G. (1997). Explaining bargaining impasse: The role of self-serving biases. *The Journal of Economic Perspectives, 11*(1), 109–126. https://doi.org/10.1257/jep.11.1.109
12. Babcock, L., Loewenstein, G., Issacharoff, S., & Camerer, C. (1995). Biased judgments of fairness in bargaining. *American Economic Review, 8*(5), 1337–1343.
13. Badger, G. J., Bickel, W. K., Giordano, L. A., Jacobs, E. A., Loewenstein, G., & Marsch, L. (2007). Altered states: The impact of immediate craving on the valuation of current and future opioids. *Journal of Health Economics, 26*(5), 865–876. https://doi.org/10.1016/j.jhealeco.2007.01.002
14. Bandura, A. (1999). Moral disengagement in the preparation of inhumanities. *Personal and Social Psychology Review, 3*, 193–209. https://doi.org/10.1207/s15327957pspr0303_3
15. Baron, J. (1995). Myside bias in thinking about abortion. *Thinking & Reasoning, 1*, 221–235. https://doi.org/10.1080/13546789508256909
16. Baumeister, R. F., & Heatherton, T. F. (1996). Self-regulation failure: An overview. *Psychological Inquiry, 7*, 1–15. https://doi.org/10.1207/s15327965pli0701_1
17. Baumhart, R. (1968). *An honest profit: What business men say about ethics in business*. Holt.
18. Becker, G. S., & Murphy, K. M. (1988). A theory of rational addiction. *The Journal of Political Economy, 96*, 675–700. https://doi.org/10.1086/261558
19. Biermann, F., & Siebenhüner, B. (2009). *Managers of global change: The influence of international environmental bureaucracies*. MIT Press. https://doi.org/10.7551/mitpress/9780262012744.001.0001
20. Bosworth, R., Cameron, T. A., & DeShazo, J. R. (2006). *Preferences for preventative public health policies with jointly estimated rates of time preference*. School of Public Health and International Affairs: North Carolina State University.
21. Bowles, S. (2004). *Microeconomics: Behavior, institutions & evolution*. Princeton University Press.
22. Cairns, J., & van der Pol, M. (2008). Valuing future private and social benefits: The discounted utility model versus hyperbolic discounting models. *Journal of Economic Psychology, 21*(2), 191–205. https://doi.org/10.1016/S0167-4870(99)00042-2

23. Camerer, C. F., Loewenstein, G., & Rabin, M. (2004). *Advances in behavioral economics*. Princeton University Press. https://doi.org/10.1515/9781400829118
24. Cameron, T. A., & Gerdes, G. R. (2003). *Eliciting individual-specific discount rates*. University of Oregon. https://doi.org/10.2139/ssrn.436524
25. Cartwright, E. (2011). *Behavioral economics*. Routledge.
26. Centeno, M. A., & Cohen, J. N. (2010). *Global capitalism*. Wiley.
27. Chabris, Ch., Laibson, D., & Schuldt, J. (2008). *Intertemporal choice*. Palgrave Dictionary of Economics.
28. Chaiken, S., Giner-Sorolla, R., & Chen, S. (1996). Beyond accuracy: Defense and impression motives in heuristic and systematic information processing. In P. M. Gollwitzer & J. A. Bargh (Eds.),*The psychology of action: Linking cognition and motivation to behavior* (pp. 553–578). Guilford Press.
29. Chapman, G. B. (1996). Expectations and preferences for sequences of health and money. *Organizational Behavior and Human Decision Processes, 67,* 59–75. https://doi.org/10.1006/obhd.1996.0065
30. Chung, S. H., & Herrnstein, R. J. (1967). Choice and delay of reinforcement. *Journal of Experimental Analysis of Behavior, 10*(1), 67–74. https://doi.org/10.1901/jeab.1967.10-67
31. Colby, A., & Damon, W. (1992). *Some do care: Contemporary lives of moral commitment*. Free Press.
32. Colinsky, J. (1996). Why bounded rationality? *Journal of Economic Literature, 34,* 669–700.
33. Coller, M., & Williams, M. B. (1999). Eliciting individual discount rates. *Experimental Economics, 2*(2), 107–127. https://doi.org/10.1023/A:1009986005690
34. Crocker, J. (1982). Biased questions in judgment of covariation studies. *Personality and Social Psychology Bulletin, 8*(2), 214–220. https://doi.org/10.1177/0146167282082005
35. Cushman, F. (2008). Crime and punishment: Distinguishing the roles of causal and intentional analyses in moral judgment. *Cognition, 108*(2), 353–380. https://doi.org/10.1016/j.cognition.2008.03.006
36. Damon, W., Menon, J., & Bronk, K. C. (2003). The development of purpose during adolescence. *Applied Developmental Science, 7*(3), 119–128. https://doi.org/10.1207/S1532480XADS0703_2
37. Darley, J. M., & Latane, B. (1968). Bystander intervention in emergencies: Diffusion of responsibility. *Journal of Personality and Social Psychology, 8,* 377–383. https://doi.org/10.1037/h0025589
38. De Woot, A. (2005). *Should Prometheus be bound? Corporate Social Responsibility*. Palgrave Macmillan. https://doi.org/10.1057/9780230502062
39. DellaVigna, St., & Malmendier, U. (2004). Contract design and self-control: Theory and evidence. *Quarterly Journal of Economics, 119*(2), 353–402. https://doi.org/10.1162/0033553041382111
40. DellaVigna, St., & Malmendier, U. (2006). Paying not to go to the gym. *American Economic Review, 96*(3), 694–719. https://doi.org/10.1257/aer.96.3.694
41. DellaVigna, St., & Paserman, M. D. (2005). Job search and impatience. *Journal of Labor Economics, 23*(3), 527–588. https://doi.org/10.1086/430286
42. Diekmann, K. A., Tenbrunsel, A. E., & Galinsky, A. D. (2003). From self-prediction to self-defeat: Behavioral forecasting, self-fulfilling prophecies, and the effect of competitive expectations. *Journal of Personality and Social Psychology, 85,* 672–683. https://doi.org/10.1037/0022-3514.85.4.672
43. Doyle, J. R. (2013). Survey of time preference, delay discounting models. *Judgment and Decision Making, 8*(2), 116–135.
44. Duflo, E., Banerjee, A., Glennerster, R., & Kothari, D. (2010). Improving immunization coverage in rural India: A clustered randomized controlled evaluation of immunization campaigns with and without incentives. *British Medical Journal, 340,* 2220. https://doi.org/10.1136/bmj.c2220

45. Duflo, E., Kremer, M., & Robinson, J. (2008). How high are rates of return to fertilizer? Evidence from field experiments in Kenya. *American Economic Review, 98*(2), 482–488. https://doi.org/10.1257/aer.98.2.482
46. Ebert, J. E., & Prelec, D. (2007). The fragility of time: Time-insensitivity and valuation of the near and far future. *Management Science, 53*(9), 1423–1438. https://doi.org/10.1287/mnsc.1060.0671
47. Epley, N., & Caruso, E. M. (2004). Egocentric ethics. *Social Justice Research, 17,* 171–187. https://doi.org/10.1023/B:SORE.0000027408.72713.45
48. Epley, N., & Dunning, D. (2000). Feeling "holier than thou": Are self-serving assessments produced by errors in self- or social prediction? *Journal of Personality and Social Psychology, 79*(6), 861–875. https://doi.org/10.1037/0022-3514.79.6.861
49. Epper, T., Fehr-Duda, H., & Bruhin, A. (2011). Viewing the future through a warped lens: Why uncertainty generates hyperbolic discounting. *Journal of Risk and Uncertainty, 43,* 169–203. https://doi.org/10.1007/s11166-011-9129-x
50. Epstein, S. (1973). The self-concept revisited. *American Psychologist, 28,* 404–416. https://doi.org/10.1037/h0034679
51. Estle, S. J., Green, L., Myerson, J., & Holt, D. D. (2007). Discounting of money and directly consumable rewards. *Psychological Science, 18,* 58–63. https://doi.org/10.1111/j.1467-9280.2007.01849.x
52. Frederick, S., Loewenstein, G., & O'Donoghue, T. (2002). Time discounting and time preference: A critical review. *Journal of Economics Literature, 40,* 351–401. https://doi.org/10.1257/jel.40.2.351
53. Fukuyama, F. (1995). *Trust: The social virtues and the creation of prosperity.* The Free Press.
54. Gardner, H. (2007). *Responsibility at work: How leading professionals act (or don't act) responsibility.* Jossey-Bass.
55. Gebert, D., & Rosenstiel, L. V. (1996). *Organisationspsychologie. Personen und Organisationen.* Kohlhammer.
56. Gentner, D. (2002). The psychology of mental models. In N. J. Smelser & P. B. Bates (Eds.), *International encyclopedia of the social and behavioral sciences* (pp. 9683–9687). Elsevier.
57. Gerson, K. (2002). Moral dilemmas, moral strategies and the transformation of gender: Lessons from two generations of work and family change. *Gender and Society, 16*(1), 8–28. https://doi.org/10.1177/0891243202016001002
58. Gigerenzer, G. (2014). *Risk savvy: How to make good decisions.* Viking.
59. Gigerenzer, G. (2016). Towards a rational theory of heuristics. In R. Frantz & L. Marsh (Eds.), *Minds, models and milieux: Commemorating the centennial of the birth of Herbert Simon* (pp. 34–59). Palgrave Macmillan.
60. Gigerenzer, G., & Gaissmaier, W. (2011). Heuristic decision making. *Annual Review of Psychology, 62,* 451–482. https://doi.org/10.1146/annurev-psych-120709-145346
61. Green, L., Fry, A. F., & Myerson, J. (1994). Discounting of delayed rewards: A life-span comparison. *Psychological Science, 5,* 33–36. https://doi.org/10.1111/j.1467-9280.1994.tb00610.x
62. Green, L., & Myerson, J. (2004). A discounting framework for choice with delayed and probabilistic rewards. *Psychological Bulletin, 130,* 769–792. https://doi.org/10.1037/0033-2909.130.5.769
63. Haidt, J. (2001). The emotional dog and its rational tail: A social intuitionist approach to moral judgment. *Psychological Review, 1089*(4), 814–834. https://doi.org/10.1037/0033-295X.108.4.814
64. Hansen, A. (2006). Do declining discount rates lead to time inconsistent economic advice? *Ecological Economics, 60,* 138–144. https://doi.org/10.1016/j.ecolecon.2006.03.007
65. Harrison, G. W., Lau, M. I., & Williams, M. B. (2002). Estimating individual discount rates in Denmark: A field experiment. *American Economic Review, 92*(5), 1606–1617. https://doi.org/10.1257/000282802762024674

66. Henderson, N., & Bateman, I. (1995). Empirical and public choice evidence for hyperbolic social discounting rates and the implications for intergenerational discounting. *Environmental & Resource Economics, 5,* 413–423. https://doi.org/10.1007/BF00691577
67. Hennigfeld, J., Pohl, M., & Tolhurst, N. (2006). *The ICCA handbook on corporate social responsibility.* Wiley.
68. Heyman, G. M. (1996). Resolving the contradictions of addiction. *Behavioral and Brain Sciences, 19,* 591–610. https://doi.org/10.1017/S0140525X00073672
69. Hoch, St. J., & Loewenstein, G. (1991). Time-inconsistent preferences and consumer self-control. *Journal of Consumer Research, 17*(4), 492–507.
70. Hornsby, J. (2007). *An empirical investigation of the effects of discounting on privacy related decisions.* Unpublished doctoral thesis, Nova Southeastern University.
71. Horowitz, J. K., & Carson, R. T. (1990). Discounting statistical lives. *Journal of Risk and Uncertainty, 3*(4), 403–413. https://doi.org/10.1007/BF00353349
72. Horton, J. J., Rand, D. G., & Zeckhauser, R. J. (2011). The online laboratory: Conducting experiments in a real labor market. *Experimental Economics, 14*(3), 399–425. https://doi.org/10.1007/s10683-011-9273-9
73. Jenkins, B. (2007). *Expanding economic opportunity: The role of large firms: Economic opportunity series.* Harvard Kennedy School.
74. Kahn, J. (2005). *The economic approach to environmental resources.* Thomson-South-Western.
75. Kahneman, D. (2011). *Thinking, fast and slow.* Farrar, Straus and Giroux.
76. Kahneman, D., Slovic, P., & Tversky, A. (1982). *Judgment under uncertainty: Heuristic and biases.* Cambridge University Press. https://doi.org/10.1017/CBO9780511809477
77. Kahneman, D., & Thaler, R. (1991). Economic analysis and the psychology of utility: Applications to compensation policy. *American Economic Review, 81,* 341–346.
78. Kahneman, D., & Tversky, A. (1974). Judgment under uncertainty: Heuristics and biases. *Science, 185*(4157), 1124–1131. https://doi.org/10.1126/science.185.4157.1124
79. Kahneman, D., & Tversky, A. (1979). Prospect theory: An analysis of decision under risk. *Econometrica, 47*(2), 263–291. https://doi.org/10.2307/1914185
80. Kahneman, D., & Tversky, A. (2000). *Choices, values, and frames.* Cambridge University Press. https://doi.org/10.1017/CBO9780511803475
81. Kant, I. (1783/1993). *Grounding for the metaphysics of morals.* Hackett.
82. Keeler, E. G., & Cretin, S. (1983). Discounting of life-saving and other non-monetary effects. *Management Science, 29,* 300–306. https://doi.org/10.1287/mnsc.29.3.300
83. Keller, L. R., & Strazzera, E. (2002). Examining predictive accuracy among discounting models. *Journal of Risk and Uncertainty, 24*(2), 143–160. https://doi.org/10.1023/A:1014067910173
84. Kirby, K. N. (1997). Bidding on the future: Evidence against normative discounting of delayed rewards. *Journal of Experimental Psychology: General, 126,* 54–70. https://doi.org/10.1037/0096-3445.126.1.54
85. Kirby, K. N., & Herrnstein, R. J. (1995). Preference reversals due to myopic discounting of delayed reward. *Psychological Science, 6*(2), 83–89. https://doi.org/10.1111/j.1467-9280.1995.tb00311.x
86. Kirby, K. N., & Marakovic, N. N. (1995). Modeling myopic decisions: Evidence for hyperbolic delay-discounting within subjects and amounts. *Organizational Behavior and Human Decision Processes, 64,* 22–30. https://doi.org/10.1006/obhd.1995.1086
87. Laibson, D. I. (1994). *Essays in hyperbolic discounting.* Unpublished doctoral thesis, MIT.
88. Laibson, D. I. (1997). Golden eggs and hyperbolic discounting. *Quarterly Journal of Economics, 11,* 443–477. https://doi.org/10.1162/003355397555253
89. Laibson, D. I., Repetto, A., & Tobacman, J. (2003). A debt puzzle. In P. Aghion, R. Frydman, J. Stiglitz, & M. Woodford, (Eds.), *Knowledge, information, and expectations in modern economics: In honor of Edmund S. Phelps* (pp. 228–266). Princeton University Press.

90. Laux, F. L., & Peck, R. M. (2007). *Economic perspectives on addiction: Hyperbolic discounting and internalities.* Social Science Research Network working paper retrieved November 3, 2015 at http://papers.ssrn.com/sol3/papers.cfm?abstract_id=1077613
91. Leary, M. R., & Forsyth, D. R. (1987). Attributions of responsibility for collective endeavors. In C. Hendrick (Ed.), *Review of personality and social psychology: Group processes* (pp. 167–188). Sage.
92. Leibovitch, A., Stremitzer, A., & Versteeg, M. (2021). *Aspirational rules.* Hebrew University of Jerusalem Legal Research Paper.
93. Lenk, H., & Maring, M. (1992). *Wirtschaft und Ethik.* Reclam.
94. Lennick, D., & Kiel, F. (2007). *Moral intelligence: Enhancing business performance and leadership success.* Wharton School.
95. Liscow, Z. D., & Markovits, D. (2022). Democratizing behavioral economics. *Yale Journal on Regulation.*
96. Loewenstein, G. (1992). The fall and rise of psychological explanations in the economics of intertemporal choice. In G. Loewenstein & J. Elster (Eds.), *Choice over time* (pp. 3–34). Sage.
97. Loewenstein, G., & Prelec, D. (1993). Preferences for sequences of outcomes. *Psychological Review, 100,* 91–108. https://doi.org/10.1037/0033-295X.100.1.91
98. Lord, C. G., Ross, L., & Lepper, M. R. (1979). Biased assimilation and attitude polarization: The effects of prior theories on subsequently considered evidence. *Journal of Personality and Social Psychology, 37,* 2098–2109. https://doi.org/10.1037/0022-3514.37.11.2098
99. Luce, R. D., & Raiffa, H. (1957). *Games and decisions: Introduction and critical survey.* Dover.
100. Luf, G. (2008). *Grundfragen der Rechtsphilosophie und Rechtsethik. Einführung in die Rechtswissenschaften und ihre Methoden.* Manz.
101. Madden, G. L., Bickel, W. K., & Jacobs, E. A. (1999). Discounting of delayed rewards in opioid-dependent outpatients: Exponential or hyperbolic discounting functions? *Experimental and Clinical Psychopharmacology, 7*(3), 284–293. https://doi.org/10.1037/1064-1297.7.3.284
102. March, J., & Simon, H. (1958). *Organizations.* Wiley.
103. Markus, H., & Wurf, E. (1987). The dynamic self-concept: A social psychological perspective. *Annual Review of Psychology, 38,* 299–337. https://doi.org/10.1146/annurev.ps.38.020187.001503
104. Marx, K. (1867/1995). *Capital: A critique of political economy.* Progress.
105. Mas-Colell, A., Whinston, M. D., & Green, J. R. (1995). *Microeconomic theory.* Oxford University Press.
106. Mastrobuoni, G., & Weinberg, M. (2009). Heterogeneity in intra-monthly consumption patterns, self-control, and savings at retirement. *American Economic Journal: Economic Policy, American Economic Association, 1*(2), 163–189.
107. Mazur, J. E. (1987). An adjusting procedure for studying delayed reinforcement. In M. M. Commons, J. A. Nevvin, & H. Rachlin (Eds.), *Quantitative analyses of behavior* (pp. 55–73). Lawrence Erlbaum.
108. McClure, S., Ericson, K., Laibson, D., Loewenstein, G., & Cohen, J. (2007). Time discounting for primary rewards. *Journal of Neuroscience, 27*(21), 5796–5804. https://doi.org/10.1523/JNEUROSCI.4246-06.2007
109. Meier, St., & Sprenger, Ch. (2010). *Stability of time preferences.* IZA Discussion Paper 4756, Institute for the Study of Labor.
110. Messick, D. M., Bloom, S., Boldizar, J. P., & Samuelson, C. D. (1985). Why we are fairer than others. *Journal of Experimental Social Psychology, 21,* 480–500. https://doi.org/10.1016/0022-1031(85)90031-9
111. Messick, D. M., & Sentis, K. P. (1979). Fairness and preference. *Journal of Experimental Social Psychology, 15,* 418–434. https://doi.org/10.1016/0022-1031(79)90047-7
112. Messick, D. M., & Sentis, K. P. (1985). Estimating social and nonsocial utility functions from ordinary data. *European Journal of Social Psychology, 15*(4), 389–399. https://doi.org/10.1002/ejsp.2420150403

113. Meyer, A. G. (2013). Estimating discount factors for public and private goods and testing competing discounting hypotheses. *Journal of Risk and Uncertainty, 46,* 133–173. https://doi.org/10.1007/s11166-013-9163-y
114. Murnighan, J. K., Cantelon, D. A., & Elyashiv, T. (2001). Bounded personal ethics and the tap dance of real estate agency. *Advances in Qualitative Organization Research, 3,* 1–40.
115. Murphy, J. G., Vuchinich, R. E., & Simpson, C. A. (2001). Delayed reward and cost discounting. *Psychological Record, 51,* 571–588.
116. Myerson, J., & Green, L. (1995). Discounting of delayed rewards: Models of individual choice. *Journal of Experimental Analysis of Behavior, 64,* 263–276. https://doi.org/10.1901/jeab.1995.64-263
117. Nelson, J. (2004). Leadership, accountability and partnership: Critical trends and issues in corporate social responsibility. *Report of the Corporate Responsibility Initiative.* Harvard University.
118. Nisbett, R., & Ross, L. (1980). *Human inference strategies and shortcomings of social judgment.* Prentice Hall.
119. O'Banion, K., & Arkowitz, H. (1977). Social anxiety and selective memory for affective information about the self. *Social Behavior and Personality, 5*(2), 321–328. https://doi.org/10.2224/sbp.1977.5.2.321
120. Okada, E. M., & Hoch, S. J. (2004). Spending time versus spending money. *Journal of Consumer Research, 31,* 313–323. https://doi.org/10.1086/422110
121. Osberg, T. M., & Shrauger, J. S. (1986). Self-prediction: Exploring the parameters of accuracy. *Journal of Personality and Social Psychology, 5,* 1044–1057. https://doi.org/10.1037/0022-3514.51.5.1044
122. Oster, S. M., & Scott-Morton, F. M. S. (2005). Behavioral biases meet the market: The case of magazine subscription prices. *Advances in Economic Analysis & Policy, 5*(1), 1–32.
123. Paese, P. W., & Yonker, R. D. (2001). Toward a better understanding of egocentric fairness judgments in negotiation. *The International Journal of Conflict Management, 12,* 114–131. https://doi.org/10.1108/eb022852
124. Petry, N. M., & Casarella, T. (1999). Excessive discounting of delayed rewards in substance abusers with gambling problems. *Drug and Alcohol Dependence, 56*(1), 25–32. https://doi.org/10.1016/S0376-8716(99)00010-1
125. Porter, M. E., & Kramer, M. R. (2003). *The competitive advantage of corporate philanthropy. Harvard Business Review on Corporate Responsibility.* Harvard Business School.
126. Porter, M. E., & van der Linde, C. (1995). Toward a new conception of the environment: Competitiveness relationships. *Journal of Economic Perspectives, 9*(4), 97–118. https://doi.org/10.1257/jep.9.4.97
127. Puaschunder, J. M. (2010). *On corporate and financial social responsibility.* Unpublished doctoral thesis, University of Vienna, Faculty of Psychology.
128. Puaschunder, J. M. (2011). *Intergenerational equity as a natural behavioral law.* Library of Congress United States of America Copyright Office Copyright Catalogue TXu001743422/2011-03-08.
129. Puaschunder, J. M. (2015a). On the social representations of intergenerational equity. *Oxford Journal of Finance and Risk Perspectives, 4*(4), 78–99.
130. Puaschunder, J. M. (2015b). The call for global responsible intergenerational leadership in the corporate world: The quest of an integration of intergenerational equity in contemporary Corporate Social Responsibility (CSR) models. In *Proceedings of the 2015 6th International Conference of the Association of Global Management Studies at Alfred Lerner Hall of Columbia University.* The Association of Global Management Studies.
131. Puaschunder, J. M. (2016a). Global responsible intergenerational leadership: The quest of an integration of intergenerational equity in Corporate Social Responsibility (CSR) models. *Annals in Social Responsibility, 2*(1), 1–12. https://doi.org/10.1108/ASR-10-2015-0016
132. Puaschunder, J. M. (2016b). Intergenerational climate change burden sharing: An economics of climate stability research agenda proposal. *Global Journal of Management and Business Research: Economics and Commerce, 16*(3), 31–38.

133. Puaschunder, J. M. (2016c). On eternal equity in the fin-de-millénaire: Rethinking capitalism for intergenerational justice. *Journal of Leadership, Accountability and Ethics, 13*(2), 11–24.
134. Puaschunder, J. M. (2016d). On the emergence, current state and future perspectives of Socially Responsible Investment (SRI). *Consilience: The Columbia University Journal of Sustainable Development, 16*(1), 38–63.
135. Puaschunder, J. M. (2016e). Socially Responsible Investment as emergent risk prevention and means to imbue trust in the post-2008/2009 world financial crisis economy. In: O. Lehner (Ed.), *Routledge handbook of social and sustainable finance* (pp. 222–238). Taylor & Francis.
136. Puaschunder, J. M. (2016f). The call for global responsible inter-generational leadership: The quest of an integration of inter-generational equity in Corporate Social Responsibility (CSR) models. In: D. Jamali (Ed.), *Comparative perspectives on global Corporate Social Responsibility* (pp. 276–289). IGI Global Advances in Business Strategy and Competitive Advantage Book Series.
137. Puaschunder, J. M. (2017a). Ethical decision making under social uncertainty: An introduction to Überethicality. *Sustainable Production and Consumption, 12,* 78–89. https://doi.org/10.1016/j.spc.2017.05.005
138. Puaschunder, J. M. (2017b). Ethical, environmental, social and governance-oriented investments. *Archives of Business Research, 5*(8), 151–159. Authorized reprint:*Pension Policy International*, August 20, 2017.
139. Puaschunder, J. M. (2017c). Financing climate justice through climate change bonds. *Oxford Journal of Finance and Risk Perspectives, 6*(3), 1–10.
140. Puaschunder, J. M. (2017d). Intergenerational leadership: Coordinating common goods and economic stability. *Journal of Applied Business and Economics, 19*(11/12), 65–78.
141. Puaschunder, J. M. (2017e). Krisenrobuste Alternativen. *Global Investor, 3*(5), 69.
142. Puaschunder, J. M. (2017f). Mapping climate in the 21st century. *Development, 59*(3), 211–216.
143. Puaschunder, J. M. (2017g). Mapping climate justice in the 21st century. *Agriculture Research and Technology, 10*(5), 1–3.
144. Puaschunder, J. M. (2018a). Climate wealth of nations. In *Proceedings of the 11th International Research Association for Interdisciplinary Studies (RAIS) Conference on Social Sciences*, Johns Hopkins University, Montgomery County Campus, Rockville, MD, United States, pp. 39–44, November 19.
145. Puaschunder, J. M. (2018b). *Corporate Social Responsibility and opportunities for sustainable financial success.* IGI Publishing.
146. Puaschunder, J. M. (2018c). Intergenerational leadership: An extension of contemporary corporate social responsibility models. *Corporate Governance & Organizational Behavior Review, 2*(1), 7–17. https://doi.org/10.22495/cgobr_v2_i1_p1
147. Puaschunder, J. M. (2018d). *Intergenerational responsibility in the 21st century.* Vernon Art and Science.
148. Puaschunder, J. M. (2018e). Nachhaltigkeit und Investment: Psychologische Aspekte von nachhaltigkeitsorientiertem Investitionsverhalten. In C. T. Schmitt & E. Bamberg (Eds.), *Psychologie und Nachhaltigkeit: Konzeptionelle Grundlagen, Anwendungsbeispiele und Zukunftsperspektiven* (pp. 127–134). Springer.
149. Puaschunder, J. M. (2019a). Mapping climate justice: Future climate wealth of nations. In J. M. Puaschunder (Ed.), *Intergenerational governance and leadership in the corporate world* (pp. 178–190). Idea Group.
150. Puaschunder, J. M. (2019b). Mental temporal accounting. In *Proceedings of the International RAIS Conference on Social Sciences and Humanities organized by Research Association for Interdisciplinary Studies (RAIS) at Johns Hopkins University* (pp. 114–124). Montgomery County Campus, Rockville, MD, USA, June 10–11.
151. Puaschunder, J. M. (2020a). *Governance and climate justice: Global South and developing nations.* Springer International & Nature/Palgrave Macmillan.
152. Puaschunder, J. M. (2020b). Long-term investments. *Encyclopedia of the UN Sustainable Development Goals: Partnerships for the Goals.*

153. Puaschunder, J. M., & Schwarz, G. (2012). *The future is now: How joint decision-making curbs hyperbolic discounting but blurs social responsibility in the intergenerational equity public policy domain.* Harvard University Situationist Law and Mind Sciences Working Paper.
154. Puaschunder, J. M. & Gelter, M. (2022). The law, economics and governance of Generation COVID-19 long-haul. *Indiana Health Law Review, 19,* 47–125.
155. Putnam, H. (2002). *The collapse of the fact-value dichotomy and other essays.* Harvard University Press.
156. Pyszczynski, T., & Greenberg, J. (1987). Toward an integration of cognitive and motivational perspectives on social inference: A biased hypothesis-testing model. *Advances in Experimental Social Psychology, 20,* 297–340.
157. Rachlin, H., Raineri, A., & Cross, D. (1991). Subjective probability and delay. *Journal of the Experimental Analysis of Behavior, 55,* 233–244. https://doi.org/10.1901/jeab.1991.55-233
158. Rand, D. G. (2012). The promise of Mechanical Turk: How online labor markets can help theorists run behavioral experiments. *Journal of Theoretical Biology, 299,* 172–179. https://doi.org/10.1016/j.jtbi.2011.03.004
159. Rawls, J. (1971). *A theory of justice.* Harvard University Press.
160. Read, D. (2001). Is time-discounting hyperbolic or subadditive? *Journal of Risk and Uncertainty, 23*(1), 5–32. https://doi.org/10.1023/A:1011198414683
161. Read, D., Frederick, S., & Airoldi, M. (2012). Four days later in Cincinnati: Longitudinal tests of intertemporal preference reversals due to hyperbolic discounting. *Acta Psychologica, 140*(2), 177–185. https://doi.org/10.1016/j.actpsy.2012.02.010
162. Read, D., Loewenstein, G., & Kalyanaraman, S. (1999). Mixing virtue and vice: The combined effects of hyperbolic discounting and diversification. *Journal of Behavioral Decision Making, 12,* 257–273. https://doi.org/10.1002/(SICI)1099-0771(199912)12:4<257::AID-BDM327>3.0.CO;2-6
163. Read, D., & van Leeuwen, B. (1998). Predicting hunger: The effects of appetite and delay on choice. *Organizational Behavior and Human Decision Processes, 76*(2), 189–205. https://doi.org/10.1006/obhd.1998.2803
164. Reed, D. D., & Martens, B. K. (2011). Temporal discounting predicts student responsiveness to exchange delays in a classroom token system. *Journal of Applied Behavior Analysis, 44,* 1–18. https://doi.org/10.1901/jaba.2011.44-1
165. Reuben, E., Sapienza, P., & Zingales, L. (2010). Time-discounting for primary and monetary rewards. *Economic Letters, 106*(2), 125–127. https://doi.org/10.1016/j.econlet.2009.10.020
166. Ross, M., & Sicoly, F. (1979). Egocentric biases in availability and attribution. *Journal of Personality and Social Psychology, 37,* 322–336. https://doi.org/10.1037/0022-3514.37.3.322
167. Rubinstein, M. (2003). Great moments in financial economics: I, present value. *Journal of Investment Management, 3*(4), 87–103.
168. Salanié, F., & Treich, N. (2005). Over-savings and hyperbolic discounting. *European Economic Review, 50*(6), 1557–1570. https://doi.org/10.1016/j.euroecorev.2005.05.003
169. Samuelson, P. (1937). A note on measurement of utility. *The Review of Economic Studies, 4,* 155–161. https://doi.org/10.2307/2967612
170. Samuelson, P., & Zeckhauser, R. (1988). Status quo bias in decision making. *Journal of Risk and Uncertainty, 1*(1), 7–59. https://doi.org/10.1007/BF00055564
171. Savage, L. J. (1954). *The foundations of statistics.* Dover.
172. Savage, L. J. (1967). Implications of personal probability for induction. *The Journal of Philosophy, 64,* 593–607. https://doi.org/10.2307/2024536
173. Schweitzer, M. E., Ordóñez, L., & Douma, B. (2004). Goal setting as a motivator of unethical behavior. *Academy of Management Journal, 47*(3), 422–432.
174. Sen, A. K. (1971). Choice functions and revealed preference. *The Review of Economic Studies, 38*(3), 307–317. https://doi.org/10.2307/2296384
175. Sen, A. K. (1977). Rational fools: A critique of the behavioral foundations of economic theory. *Philosophy & Public Affairs, 6*(4), 317–344.
176. Sen, A. K. (1993). Internal consistency of choice. *Econometrica, 61*(3), 495–521. https://doi.org/10.2307/2951715

177. Sen, A. K. (1997). Maximization and the act of choice. *Econometrica, 65*(4), 745–780. https://doi.org/10.2307/2171939
178. Sen, A. K. (2002a). Consistency of choice. In A. Sen (Ed.), *Rationality and freedom* (pp. 468–498). Harvard University Press.
179. Sen, A. K. (2002b). Goals, commitment, and identity. In A. Sen (Ed.), *Rationality and freedom* (pp. 15–32). Harvard University Press.
180. Shackle, G. L. S. (1955). *Uncertainty in economics and other reflections*. Cambridge University Press.
181. Shah, A. K., & Oppenheimer, D. (2008). Heuristics made easy: An effort-reduction framework. *Psychological Bulletin, 134*(2), 207–222. https://doi.org/10.1037/0033-2909.134.2.207
182. Shefrin, H., & Thaler, R. H. (1988). The behavioral life-circle hypothesis. *Economic Inquiry, 26,* 609–643. https://doi.org/10.1111/j.1465-7295.1988.tb01520.x
183. Sherman, S. J. (1980). On the self-erasing nature of errors of prediction. *Journal of Personality and Social Psychology, 39,* 211–221. https://doi.org/10.1037/0022-3514.39.2.211
184. Shui, H., & Ausubel, L. M. (2004). *Time inconsistency in the credit card market*. Social Science Research Network working paper retrieved October 25, 2015 at http://ssrn.com/abstract=586622
185. Sichler, R. (2003). Neue Arbeitswelt und Autonomie. Aspekte menschlicher Selbstbestimmung im flexibilisierten Kapitalismus. In A. Birnbaumer & G. Steinhardt (Eds.), *Der flexibilisierte Mensch. Subjektivität und Solidarität im Wandel* (pp. 139–149). Asanger.
186. Sichler, R. (2004). Autonomie. Ihre Konstitution und Ambivalenz im Kontext von Arbeit und Lebensführung. In W. G. Weber, P.-P. Pasqualoni, & Ch. Burtscher (Eds.), *Wirtschaft, Demokratie und soziale Verantwortung. Psychologie und Beruf* (pp. 104–126). Vandenhoeck & Ruprecht.
187. Simon, H. A. (1957). *Models of man: Social and rational*. Wiley.
188. Simon, H. A. (1979). Rational decision making in business organizations. *American Economic Review, 69*(4), 493–513.
189. Simon, H. A. (1983). *Reasons in human affairs*. Stanford University Press.
190. Simon, H. A., & Bartel, R. D. (1986). The failure of armchair economics. *Challenge, 29*(5), 18–25. https://doi.org/10.1080/05775132.1986.11471113
191. Slonim, R., Carlson, J., & Bettinger, E. (2007). Possession and discounting behavior. *Economic Letters, 97*(3), 215–221. https://doi.org/10.1016/j.econlet.2007.03.018
192. Smith, A. (1759/1976). *The theory of moral sentiments*. Oxford University Press.
193. Snyder, C. R., & Swann, W. B. (1978). Behavioral confirmation in social interaction from social perception to social reality. *Journal of Experimental Social Psychology, 14,* 148–162. https://doi.org/10.1016/0022-1031(78)90021-5
194. Sterner, T. (1994). Discounting in a world of limited growth. *Environmental & Resource Economics, 4,* 527–534. https://doi.org/10.1007/BF00691927
195. Strotz, R. H. (1956). Myopia and inconsistency in dynamic utility maximization. *Review of Economic Studies, 23,* 165–180. https://doi.org/10.2307/2295722
196. Suri, W., & Watts, D. J. (2011). Cooperation and contagion in web-based, networked public goods experiments. *Public Library of Science, 6*(3), 16836.
197. Taylor, S. E. (1989). *Positive illusions: Creative self-deception and the healthy mind*. Basic Books.
198. Tenbrunsel, A. E., & Messick, D. M. (2004). Ethical fading: The role of self-deception in unethical behavior. *Social Justice Research, 17*(2), 223–236. https://doi.org/10.1023/B:SORE.0000027411.35832.53
199. Thaler, R. H. (1980). Toward a positive theory of consumer choice. *Journal of Economic Behavior & Organization, 1,* 39–80. https://doi.org/10.1016/0167-2681(80)90051-7
200. Thaler, R. H. (1981). Some empirical evidence on dynamic inconsistency. *Economics Letters, 8*(3), 201–207. https://doi.org/10.1016/0165-1765(81)90067-7
201. Thaler, R. H. (1999). Mental accounting matters. *Decision Making, 12,* 183–206.
202. Thaler, R. H., & Shefrin, H. M. (1981). An economic theory of self-control. *Journal of Political Economy, 89,* 392–406. https://doi.org/10.1086/260971

203. Thaler, R. H., & Sunstein, C. (2008). *Nudge: Improving decisions about health, wealth, and happiness*. Yale University Press.
204. Thompson, L., & Loewenstein, G. (1992). Egocentric interpretations of fairness and interpersonal conflict. *Organizational Behavior and Human Decision Processes, 51,* 176–197. https://doi.org/10.1016/0749-5978(92)90010-5
205. Trivers, R. (1971). The evolution of reciprocal altruism. *Quarterly Review of Biology, 46,* 35–57. https://doi.org/10.1086/406755
206. Trope, Y., & Fishbach, A. (2000). Counteractive self-control in overcoming temptation. *Journal of Personality and Social Psychology, 79*(4), 493–506. https://doi.org/10.1037/0022-3514.79.4.493
207. Tversky, A., & Kahneman, D. (1974). Judgment under uncertainty: Heuristics and biases. *Science, 185,* 1124–1131. https://doi.org/10.1126/science.185.4157.1124
208. Vallone, R. P., Griffin, D. W., Lin, S., & Ross, L. (1990). Overconfident prediction of future actions and outcomes by self and others. *Journal of Personality and Social Psychology, 58,* 582–592. https://doi.org/10.1037/0022-3514.58.4.582
209. van der Pol, M., & Cairns, J. (2001). Estimating time preferences for health using discrete choice experiments. *Social Science and Medicine, 52*(9), 1459–1470. https://doi.org/10.1016/S0277-9536(00)00256-2
210. Wade-Benzoni, K. A. (1996). *Intergenerational discounting*. Unpublished doctoral thesis, Northwestern University, Evanston, IL.
211. Warner, J. T., & Pleeter, S. (2007). The personal discount rate: Evidence from military downsizing programs. *American Economic Review, 91*(1), 33–53. https://doi.org/10.1257/aer.91.1.33
212. Wertenbroch, K. (1998). Consumption self-control by rationing purchase quantities of virtue and vice. *Marketing Science, 17*(4), 317–337. https://doi.org/10.1287/mksc.17.4.317
213. Winston, G. C. (1980). Addiction and backsliding: A theory of compulsive consumption. *Journal of Economic Behavior & Organization, 1,* 295–324. https://doi.org/10.1016/0167-2681(80)90009-8
214. World Bank. (2015). Green bonds attract private sector climate finance. *World Bank Brief*.
215. World Development Report. (2015). *Mind, society and behavior*. World Bank.
216. Zauberman, G., Kim, B. K., Malkoc, S. A., & Bettman, J. R. (2009). Discounting time and time discounting: Subjective time perception and intertemporal preferences. *Journal of Marketing Research, 46*(8), 543–556. https://doi.org/10.1509/jmkr.46.4.543
217. Zuroff, D. C., Colussy, S. A., & Wielgus, M. S. (1983). Selective memory and depression: A cautionary note concerning response bias. *Cognitive Therapy and Research, 7*(3), 223–232. https://doi.org/10.1007/BF01205136

Part II
Digital Behavioral Economics

2 Communication in the Twenty-First Century

One of the most novel implications of heuristics, biases, and nudges addresses behavioral finance—a growing field concerned with how to improve financial well-being through the sound understanding of how people actually behave in markets. Contemporary economic growth theories and financial well-being advice are connected to digitalization. The emergence of digital markets has also risen novel forms of digital inequality. While digitalization and inequality are predominant features of our times, hardly any information exists on inequality inherent in digitalization and underlying any innovative change. This chapter breaks novel ground in elucidating digital inequality from an behavior economics perspective. Understanding digital inequality helps explaining some behavioral finance techniques that can be used to enhance your financial gain by diversification, investing in crises-robust, long-term sustainable market options, demographics-based forecasting, saving money through tangibility and safe havens, or reaping benefit from outperforming market strategies. Particular emphasis will be given to investigating the role of information and communication for market reactions but also social influences in financial market management.

We live in the age of digitalization. Digital disruption is the advancement of our lifetimes. Never before in the history of humankind have human beings given up as much decision-making autonomy as today to a growing body of artificial intelligence (AI) and robotics entering our daily lives. Algorithms and big data are shaping society as never before in the history of humankind. Digitalization features a wave of self-learning entities that generate information from exponentially growing big data sources that are encroaching every aspect of our daily lives. Inequality is the most significantly pressing concern of our times. Ample evidence exists in economics, law and historical studies that multiple levels of inequality dominate the current socio-dynamics, politics and living conditions around the world. Social inequality stretches from within nation states to global dimensions but also intergenerational inequality levels. While digitalization and inequality are predominant features of our times, hardly any information exists on inequality inherent in

J. Puaschunder, *Advances in Behavioral Economics and Finance Leadership*, Contributions to Economics, https://doi.org/10.1007/978-3-031-15710-3_2

digitalization. The following chapter will therefore break new ground in theoretically arguing for inequality being an overlooked by-product of innovative change of modern capitalism. A multi-faceted analysis will draw a contemporary digital inequality account from a behavior economics perspective.

This chapter is designed to aid academics and practitioners in understanding the advantages but also the potential inequalities imbued in digitalization. It sets a historic landmark to capture the Zeitgeist of our digitalization disruption heralding unexpected inequalities stemming from innovative change. Understanding the contemporary digitalization disruption whole-roundedly in its advantageous innovation capacities but also potential societal, international and intertemporal unequal gains and losses perspectives will help navigating through economic changes wisely and strategically. The rising digital inequality gap will be portrayed as a potential and hidden source of societal and international tensions as well as intergenerational imbalances, which can be overcome with wise individual choices, strategic market behavior and well-managed public policy advancements.

Overall, this chapter also targets at proposing solutions for those who strive to take advantage of digitalization innovation and work to improve the concept of digitalization market disruptions. Interdisciplinary and heterodox economic viewpoints will be coupled with social justice equality pledges and the sustainable development goals, which will bestow the reader with a long-term future-oriented socially-conscientious focus. A broad readership that has leaders from across all sectors is targeted with this approach. The core target audience are scholars and practitioners of tomorrow, given the timely topics addressed. The future outlooks may aid practitioners of finance, governance, public policy, and international development leadership to alleviate the most pressing contemporary societal predicaments around digitalization.

2.1 Nudging and Winking in the Digital Era

As a novel application of political economy, the behavioral insights approach appears to be limited and hold unforeseen risks of social class division in the nudgital society [2, 27, 31]. While the motivation behind nudging appears as a noble endeavor to foster peoples' lives around the world in very many different applications [17], the nudging approach raises questions of social hierarchy and class division. The motivating force of the nudgital society may open a gate of exploitation of the populace and—based on privacy infringements—stripping them involuntarily from their own decision power in the shadow of legally permitted libertarian paternalism and under the cloak of the noble goal of welfare-improving global governance. Nudging enables nudgers to plunder the simple uneducated citizen, who is neither aware of the nudging strategies nor able to oversee the tactics used by the nudgers. The nudgers are thereby legally protected by democratically-assigned positions they hold or by outsourcing strategies used, in which social media plays a crucial rule.

Digitalization advances human interaction and international access to resources such as digital workplaces, e-healthcare and remote education. The worldwide rising data transfer and drastic shift to online marketplaces during the COVID-19 lockdowns exacerbated digitalization. But markets flourishing in online virtual spaces also widened inequality gaps in the digital world. In order to cope with a massively-rising amount of information transferred online in a fastening speed, online searchplaces, such as Google, Yahoo, Bing, DuckDuckGo and Yandex, are mainly maintained by constant algorithmic activity and artificially-intelligent choices what information to display.

In the digital age, nudging in competitive areas can also lead to nudges with unpleasant information being used in competition against people. For example, online search engine algorithms can be strategically manipulated against humans in competitive situations. Searching for them on the Internet can maliciously be negatively impacted by erasing information online by flagging appropriate content as inappropriate or hyping broken links, outdated information and disinformation with clickfarms.

As online market places have gained prominence for screening and scanning candidates in workplace-related contexts in the most recent decade, the quest for creating fair online spaces has reached unprecedented momentum. Despite involuntary replications of existing biases and stereotypes on the internet, online searchplaces also appear to have become competitively-contested terrains that are strategically manipulatable. While we have a most important and emerging stream of literature on algorithmic discrimination and inherent biases in searchplaces and internet social forums; less is known and talked about strategic searchplace distortion. Often unnoticingly, misinformation, disinformation, spamming and erasure from searchplace results are thereby being used in a competitive sense employing Search Engine Optimization (SEO) in a negative and harmful way.

The anonymity of the Internet, but also the subtle ways in which perception is manipulated, are unique innovations in the digital age. Global governance institutions and public administration as well as private companies are currently recognizing this new type of danger of searchplace discrimination online. More power must be given to individuals to reclaim their searchplace space and communication channels enacted to adjust distorted searchplace representations.

To alleviate this problem, legal and regulatory instruments as well as voluntary industry measures and community standards raising awareness and discrimination protection are needed. In the digital age, social media revolutionized human communication around the globe, yet also opened opportunities to unprecedentedly reap benefits from information sharing and big data generation. The law of motion of the nudging societies holds an unequal concentration of power of those who have access to compiled data and who abuse their position under the cloak of hidden persuasion and in the shadow of paternalism. In the nudgital society, information, education, and differing social classes determine who the nudgers and who the nudged are. Humans end in different silos or bubbles that determine who has power and control and who is deceived and being ruled. The owners of the means of governance are able to reap a surplus value in a hidden persuasion, protected by the legal vacuum to curb libertarian paternalism, in the moral shadow of the unnoticeable guidance and under the cloak of the presumption that some know what is more rational than others [5]. All these features lead to an unprecedented contemporary class struggle between the nudgers (those who nudge) and the nudged (those who are nudged), who are divided by the implicit means of governance in the digital scenery.

2.1.1 On the Collective Soul of Booms and Busts

Globalization led to an intricate set of interactive relationships between individuals, organizations, and states [7]. Deepening nets of interactions challenge human foresight [11]. Collective interaction effects lead to hard-to-foresee fallacy of composition societal downfalls [26]. As complex interdependencies may hold unknown outcomes for society, highly integrated international communities are under the whim of unexpected socio-economic developments. In seeking to shed light onto implicit system failures' socio-economic consequences down the road and potentially disastrous outcomes of cumulative actions triggering mass movements; the "emergent" risk theory outlines unexpected dangers and insufficiently-described shadows of the invisible hand of the world economy in the age of globalization [8, 18, 26].

Since the post-World War period, the world globalized. International economic activities now involve a larger number of countries and sectors than at any time in history and reaches deeper into every human life than ever [12]. Global interaction possibilities have also made communication unprecedentedly complex. With growing globalization and quickening of transfer speed, information flows may impose unknown systemic economic risks on a global scale [6, 20]. Nowadays, information flow has no longer limited local effects but potentially unforeseen global consequences [15, 29, 30]. This part of the book will argue that an asymmetry of information may create risks within economic markets for fueling booms and downturns. The role of new social online media platforms for financial markets will be explored.

While by the end of the 1960s the most renowned economists agreed that recession was preventable, history proved them wrong [4]. For instance, the period from 1940 to 1973 became renowned for a time of economic prosperity, from 1973 a worldwide recession set in. Historic post-World War II booms were transitioned to downturn from the mid-1970s on as the economic performance declined in the industrialized world. While in the 1980s and 1990s the economy seemed to expand again, from 2000 on productivity slowed, again, from 1998 to 1999 stock markets and currencies crashed or halted in 2001 and 2008 as well as during the current post-Coronavirus recession. Today, supply side theory explains the downturn dependent on pressures from labor [4]. Behavioral macro-economists give credit to the unplanned, uncoordinated, and competitive nature of capitalist production as well as the problem of aggregates deviating from the individual's choice predictions.

Economic indicators of product wages, international competition, output-capital ratio, and post-tax profits have been studied extensively to derive conclusions about economic pre-indicators of crises and recommendations for improving economic systems. The role of inflation is widely known in inverse relation to unemployment to determine economic conditions [1, 3]. The international transmission of inflation is discussed in historic examples [28]. Adaptive expectations are built by information about the rate of inflation based on current and past experience with inflation [28]. Disequilibrium inflation occurs when the actual rate of inflation is greater than the expected and people try to realize the real income increase followed by social unrest and industrial conflict [28].

While business cycle theories primarily focus on describing economic correlates of booms and busts such as tight labor markets, investment trends [3], and the uneven development throughout the world causing advantages and disadvantages in the economic impact of booms and busts around the world [3], less attention is shed on socio-economic correlates that build expectations leading to irrational exuberance. Yet information about the economy is key in shaping ideas and intentions of individual market actors. Communication on markets is thus argued to be underlying long-term economic trends as well. For instance, information transfer is essential for innovation and markets to pick up new ideas. Information shaping expectations are the basis of investment trends. The collective mood in society shapes very many investment decisions amalgamating into economic trends that determine economic dynamism. Individual communication may also be the basis of social unrest and waves of strikes, which have been shown to be an underlying factor determining wages and rates of profit for capitalist [3]. Access to information may also play a key role in price comparisons around the world, which are the basis of outsourcing and capital allocation decisions [3]. Information portrayed in media may also determine the investment mood and credit liquidity preferences of individuals and institutional representatives. All these correlates set the tact in determining the booms and busts of markets as well as the long-term cycles.

The role of information and individual communication for economic long-term cycles, however, has been—so far—overlooked. Communication interventions are neglected in a wealth of writings on Federal Reserve and Central Bank macroeconomic interventions ranging from lowering interest rates to direct monetary stimulus such as quantitative easing. Studying the effect of information and communication on economic correlates to shape market decisions may offer invaluable novel insights on how bubbles start and economic fluctuations can be smoothed.

Through capturing the interplay of communication and the economy, this part is meant to shed light on economic downfalls in a heterodox fashion in order to serve as a window of opportunity for alleviating negative externalities of emergent risks of globalization. This part therefore proposes to investigate the unprecedentedly -described role of information in building and fueling economic booms and downturns with particular attention to contemporary digitalization trends. Theoretically, economic studies of the nature of economic booms and busts aid in pursuing the greater goal of deriving recommendations on how to stabilize economic markets in the instant communication century and finding an optimum balance of deregulated market systems and governmental control.

The chapter is organized as follows: An introduction to the history of economic cycles leads to the analysis of the role of information in the creating of economic booms and busts. The inherent potential of inequality in innovation's creative destruction will be outlined by the example of digitalization inequality. Recommendations on how to create more stable economic systems by the strategic use of nudges avoiding emergent risks within economic market systems and bridging inequality gaps within society are given in the discussion followed by a conclusion and prospective future research outlook.

2.1.2 Nudgital: Critique of Behavioral Political Economy

In order to understand the impetus of the nudgital society, we need to study the laws of governance in the age of information. From primitive communication between human beings, a civilization of information transfer with centralized state authority and market value in communication control has emerged but also competitive notions regarding searchplace information display. In the twenty-first century, the turnover of information and the aggregation of social informational capital has revolutionized the world. In the wake of the emergence of new social media communication and interaction methods, a facilitation of the extraction of surplus value in shared information has begun. In the following, the main ideas behind the social media marketplace are dissected in order to show how surplus value through access to amalgamated information over distance and time is realized and an implicit social class division between the nudgers and the nudged evolved in the digital age.

Imagine signing up for a free social media tool, such as Facebook, Instagram, or Twitter. You will connect with other people and constantly upload information about yourself, your life, and your friends in order to share and benefit from shared information. New media online communication tools have made the world flat. No social hierarchies exist when considering one can follow powerful peoples' news on Twitter and the opportunity to connect and feedback influential individuals' web appearances. No distance in space and time seem to matter, when considering our opportunities to instant messaging around the globe 24/7, send messages posthumously, and compile massive amounts of big data on a constant basis, which can be stored eternally. All these information flows can be combined to find fascinating behavioral insights and learn about market trends, thanks to unprecedented computational power in the twenty-first century. Computational procedures for data collection, storage, and access in the large-scale data processing have been refined for real time and historical data analysis, spatial and temporal results as well as forecasting and nowcasting in the last decade. All these advancements have offered a multitude of in-depth information on human biases and imperfections as well as social representations and collective economic trends [19, 20, 23, 32, 33].

But are these features of the digital age solely positive advancements of humankind or do they hold problematic emergent risks for humanity and implicit danger of societal stratification [8]? Are the behavioral insights gravitating toward an elite that holds the power to reap benefits from the many who innocently share personal information by giving into the humane-imbued need for communication? Can the website be manipulated in a targeted manner and who regulates the Internet search functions when it comes to competition and anonymous postings or so-called click farms, i.e., purchased Internet requests that change the search results competitively, strategically or even unethically? Do the novel computational power advantages lead to a divided society and an unequal distribution of political power and means to steer collective action [9]?

This part of the book examines the relationship between heuristics, nudging, unfair searchplace discrimination and social class in the digital age. Thereby, the chapter argues that the strategic use of heuristics differs across social classes. Nudging is shown to have become a prerogative of the elite, who has more information given a difference in access to compiled information.

In the nudgital society, information about others plays a key role in determining a competitive advantage. In its most cruel sense, nudges can be used in a targeted manner in competitive situations, for example when Internet search engines manipulate results in a targeted manner. For example, unfavorable comments, misinformation or even disinformation can be preceded by click farms in Internet searches in order to associate people with unfavorable Internet appearances in an unpleasant way.

The digital age has also brought about unprecedented opportunities to amalgamate big data information that can directly be used to derive inferences about people's preferences in order to nudge and wink them in the nudgitalist's favor. Social classes have different levels of education and insights about the nudgital act, which lead to different confidence levels in their economic choices to act on the nudgital insights and to abstain from opt-out devices. Those who reap surplus value are naturally blessed with higher income levels and elevated educational backgrounds coupled with self-confidence, which leads to less susceptibility to fall for nudges and winks. These elite circles are more confident in their decision-making and respond more well-informed to opt-out options.

In today's nudgital society, information has become a source of competitive advantage. Technological advancement and the social media revolution have increased the production of surplus value through access to the combined information. Human decisions to voluntarily share information with others in the search for the humane pleasure derived from communication are objectified in human economic relations. Unprecedented data storage possibilities and computational power in the digital age have leveraged information sharing and personal data into an exclusive asset that divides society into those who have behavioral insights derived from a large amount of data (the nudgers) and those whose will is manipulated (the nudged).

The implicit institutional configuration of a hidden hierarchy of the nudgital society is structured as follows: Different actors engage in concerted action in the social media marketplace. The nudgital brokers are owners and buyers of social media space, which becomes the implicit means of production. In the age of instant global information transfer, the so-called social media industrialist-capitalist provides the social media platform, on which the social media consumer-workers get to share information about their life and express their opinion online for free. In their zest for a creation of a digital identity on social media platforms, a "commodification of the self" occurs. Social media consumer-producer-worker are sharing information and expressing themselves, which contributes to the creation of social media experience.

The hidden power in the nudgitalist society is distributed unevenly, whereby the social media consumer-workers are falling for their own human nature to communicate with one another. Social media consumer-workers also engage in social media expression as for their social status striving in the social media platforms, where they can promote themselves. By posing to others in search of social status enhancement and likes, they engage in voluntary obedience to the social media capitalist-industrialist who sells their labor power product of aggregated information to either capitalists or technocrats. The social media consumer-worker's use value is inherent in their intrinsic motivation to satisfy a human need or want to communicate and gain respect from their community. The use value of the commodity is a social use value, which has a generally accepted use value derived from others' attention and respect in the wake of information sharing in society. The social media provider gives the use value an outlet or frame, which allows the social media consumer-worker to express information, compare oneself to others, and gain information about the social relation to others. The consumer-laborer thereby becomes the producer of information, releasing it to the wider audience and the social media industrialist. The tool becomes an encyclopedic knowledge and joy source derived from the commodity. This use value only becomes a reality by the use or consumption of the social media and constitutes the substance of consumption. But the use of social media is not an end in itself but a means for gathering more information that can then be amalgamated by the social media capitalist-industrialist, who harvests its use value to aid nudgers. It is a social form of wealth, in the form of social status and access to knowledge about others that the use value materializes on the side of the industrialist in the exchange value. For the social media industrialist, who is engaged in economic and governmental relations, the exchange value of the information provided by his or her social media consumer laborers is the information released and consumption patterns studied. In exchange, this allows us to derive knowledge about purchasing and consumption patterns of the populace and therefore creates opportunities to better nudge consumers and control the populace. With the amassed information, the social media industrialist-capitalist can gain information about common trends that can aid governmental officials and technocrats in ensuring security and governance purposes. Further, the social media platform can be used for marketing and governmental information disclaimers. Exchange value is a social process of self-interested economic actors taking advantage of information sharing based on utility derived from consuming the social media. The social media industrialist-capitalist can negotiate a price based on the access to the social media consumer-worker's attention and sell promotion space to marketers. The exchange value of the commodity of information share also derives from the subjective perception of the value of amalgamated data. Exchanged information can be amalgamated by the social media industrialist-capitalist and traded to other market actors. Exchange value is derived from integrating everything the worker is and does, so both in his or her creative potential and how he or she relates to others. Not just labor power but the whole person becomes the exchange value, so one could even define the consumer-worker as utility-provider. The amalgam of information serves as indicator how the majority reacts to changing environments, which allows inferences about current trends and predictions about the future. Exchange value also stems from the exchange of the commodity of amalgamated information that enables an elite to nudge the general populace. Underlying motives may be the desire for prestige and distinction on both sides—the industrialist-capitalist's and the consumer-worker's. On the industrialist-capitalist's side monetary motives may play a role in the materialization of information; on the consumer-worker's side, it is the prestige gained from likes, hence, respect for an online identity created. The benefits of the superior class are the power to nudge, grounded on the people's striving for prestige and image boosts. Emotions may play a vital role in seducing people to share information about themselves and derive pleasure for sharing [10, 14, 16]. Social norms and herding behavior may be additional information sharing drivers [21]. The realization of prestige stems from creating a favorable image of oneself online, which signs up the workers in a psychological quasi-contract to provide more and more information online and in a self-expanding value. Prestige is also gained in the materialization of information as asset by the capitalist-industrialist, who reaps the surplus value of the commodification of the self of the consumer-worker. The social media capitalist-industrialist therefore increases their capital based on the social media consumer-worker's innocent private information share. The social media capitalist-industrialist also accumulated nudgital, the power to nudge.

This information sharing opens a gate for the social media provider to reap surplus value from the information gathered on social platforms and to nudge the social media consumer-producers or resell their amalgamated information to nudgers. Crucial to the idea of exploitation is the wealth or power of information in the digital age. Surplus of information can be used to nudge in markets and by the force of governments. To acknowledge social media consumers as producers leads to the conclusion of them being underpaid workers in a direct wage-labor exploitation. Surplus gravitates toward the social media owning class. Information becomes a commodity and commodification a social product by the nature of communication. Commodification of information occurs through the trade of information about the consumer-worker and by gaining access to nudge consumer-workers on social platforms.

The transformation of a labor-product into a commodity occurs if information is used for marketing or governance purposes to nudge people. In the contemporary big data society, the nudged social media user therefore ends up in a situation where they are unwaged laborers, providing the content of entertainment within social media, whereas the social media industrialist-capitalist, who only offers the information brokerage platform, reaps benefits from the amalgamated information shared.

Overall, there is a decisive social role difference between the new media capitalist-industrialist and the social media consumer-worker. The social media provider is an industrialist and social connection owner, who lends out a tool for people to connect and engage with. As the innovative entrepreneur who offers a new media tool, the industrialist also becomes the wholesale merchant in selling market space to advertisement and trading information of his customers or workers, who are actively and voluntarily engaging in media tools [25]. The social media consumers turn into workers, who are even missing direct monetary remuneration for their information share since being engaged in the new media tool rather than selling their labor power for money in the marketplace holds opportunity costs of foregone labor. While selling their commodity labor power, the social media consumer-workers are also consumers of the new media tool laden information, which can be infiltrated with advertisement. The social media capitalist-industrialist not only reaps exchange value benefits through access to people's attention through selling advertisement space, but also grants means to nudge the consumers into purchasing acts or wink the populace for governance authorities. The social media capitalist-industrialist thereby engages in conversion of surplus value through information sharing into profit as well as selling attention space access and private data of the consumer-workers. When the new media consumer-workers' amalgam of provided information gets added up to big datasets, it can be used by capitalists and governance specialists. Over time the nudgital society emerges, as the nudging social media industrialist-capitalists form a Gestalt of several bits and pieces put together about the nudged social media consumer-producer-workers. Information gets systematically added up providing invaluable behavioral insights. Information in its raw form and in amalgamated consistency then gets channeled from the broad working body on social media into the hands of a restricted group or societal class. This circulation of information and the distribution into those who provide a medium of information exchange and those who exchange information then leads to an inherent social class divide of society into those who nudge and those who are nudged.

The technological complexity of digital media indicates how interrelated social, use, and exchange value creation are. All commodities are social products of labor, created and exchanged by a community, with each commodity producer contributing his or her time to the societal division of labor. Use value is derived by the consumer-worker being socially related insofar as private consumption becomes collective. The use value thereby becomes the object of satisfaction of the human need for social care and want for social interaction. The use value is modified by the modern relations of production in the social media space as the consumer-worker intervenes to modify information. What the consumer-worker says on social media, for the sake of communication and expression but also in search for social feedback, is confined by the social media industrialist-capitalist, who transforms the use value into exchange value by materializing the voluntary information share by summing it up and presenting it to nudgers, who then derive from the information marketability and nudgitability of the consumer-workers. All information sharing has value, or labor value, the abstract labor time needed to produce it. The commodification of a good and service often involves a considerable practical accomplishment in trade. Exchange value manifests itself totally independent of use value. Exchange means the quantification of data, hence putting it into monetary units. In absolute terms, exchange value can be measured in the monetary prices social media industrialist-capitalists gain from selling advertisement space to nudging marketers but also to public and private actors who want to learn about consumer behavior in the digital market arena and influence consumers and the populace.

The exchange value can also be quantified in the average consumption-labor hours of the consumers-workers. While in the practical sense, prices are usually referred to in labor hours, as units of account, there are hidden costs and risks that have to be factored into the equation, such as, for instance, missing governmental oversight and taxing of exchange value.

In the nature of exchange, nudgital becomes an abstract social power, a property claim to surplus value through information. Value can be expropriated through the exchange of information between the industrialist-capitalist and the nudgitalist. Exchange value has an inherent nature of implicit class division. Exchange value represents the nudgitalists' purchasing power expressed in his or her ability to gain labor time that is required for information sharing as a result of the labor done to produce it and the ability to engage in privacy infringements. The social media industrialist-capitalists implicitly commands labor to produce more of data through social nudging and tapping into humane needs to communicate and express themselves, whereby he or she uses a reacting army of labor encouraging information share through social gratification in the form of likes and emoticons. The reacting army of labor is comprised of social media users, who degrade into hidden laborers that are not directly compensated for their information share and cheerleading others to do the same. The nudgital society's paradox is that information sharing in the social compound gets pitted against privacy-protecting alienation.

The social relations of production in the social media domain and existing within economic exchange of big data are yet rather uncaptured. The social concept of information value therefore needs to be highlighted in order to derive conclusions about the labor theory of social media exchange value. While social media appears to create a more egalitarian society, and social hierarchies have indeed become flat in the availability to connect with different social strata around the world on an instant basis, a domination occurs in human society through the nudgers, who gain access to private information of social media users.

In the amalgamation of data individuals' private information allows us to predict trends but also to manipulate the consumers and populace. While for the consumer-worker information sharing seems no concern since it is with his/her preferred circles, the industrialist-capitalist gains an elevated position through the exchange value, leveraging him into a quasi-bourgeois class, thanks to the voluntary information share of his or her workers.

Nudgers and nudged form different social classes. The nudgers are those who augmented a higher than average amount of information value in society, while simultaneously diminishing the privacy and economic value creation of the nudged. Decision-making biases and heuristics come to play to create illusions in order to maximize economic value. The implicit governance system of the nudgital society continues to operate behind the backs of the nudgers and the nudged as the nudgers gain big data information over time and different media spaces.

2.1.3 The Nudging Divide in the Twenty-First Century

The insights gained about the nudgital society lead to the demand to rewrite economics. Standard neoclassical economic theories do not hold, when it comes to the nudgital society. In always striving to increase his or her power, the social media capitalist-industrialist constantly seeks to accumulate power in the form of information about others in order to use this information to reap exchange value through selling the amalgam of the information to create nudges for capitalists and technocrats. Since the social media industrialist has no direct contract with social media consumer-producers, he or she cannot coerce or enforce work discipline rather than having to entertain the workforce. Social media capitalist-industrialists therefore constantly have a need to find novel ways to entertain their online entourage workforce to stay online and amass information. The marginal utility of consumption rate is—contrary to neoclassical economic theory—increasing as the more time one spends on social media, the more pleasure is derivable as for making more connections and having a more complex network with more access to information of a wider variety of people.

Contrary to the falling rate of profit concept prevailing in economic models [25], the rate of profit for the nudgital society could increase over time and with claiming nudging space. The rate of profit is rising with the more people being engaged in the network and the capitalist-industrialist's successful efforts to restlessly and insatiable accumulate information. The more the people join a network, the more time they may spend on the social media tool and the more likely they are assumed to release information and voluntarily share information. So contrary to classical economic market models, the rate of profit for the new media industrialist-capitalist is assumed to be rising with the more people engaging in his or her market tool. Yet, there is a tendency of the rate of profit to fall if other social media contestants invent other social media tools that distract workers from their task of sharing information. The industrialist-capitalist is thereby in a constant battle with other social media providers for the attention of customer-workers. Since the customer-workers are non-financially rewarded, their attention has to be drawn by the industrialist-capitalist, who only technically intervenes, not actively contributes information. The capitalist-industrialist is under constant pressure of the market needing to track the wants and needs of consumers and keeping them motivated to engage on social media and share information in order to collect information of individuals throughout all social strata of society. This process may not only be influenced by economic and technical factors but also socio-political and cultural ones, insofar as it involves property rights, claims to private resources, and consumes time while being at risk to infringe upon quality and safety of use. In addition, negative market forces are fake news and alternative facts. Alternative facts can curb people's motivation to engage in social media and spend time on certain social networks.

The falling rate of profit in the nudgital world could also be decreasing if people start getting bored by social media and not upload information as they used to. This leads to a constant struggle for new social media tools and entertainment features where to derive novel utility or expanded utility from. Novel and newly designed systematic encouragements (e.g., the like button) and the development of technological capabilities of all kinds become an integral consequence for the circuit of information accumulation. The constant need to create surplus value and to protect oneself against forces that erode information sharing is alleviated by technological innovations. When innovation takes the form of a new product, the capitalist enjoys a monopolistic profit advantage, which yet may often be short-lived. Novel surplus may derive from activities on social media that bring unique novel pleasure or use values that yield a profit to their organizers in the institutional structure of the system and its technical apparatus.

The state of the market determines the social media capitalist-industrialist's impetus. The social media capitalist-industrialist's productive activity is an ongoing process determined by attention people give to social market tools. The market reality is the conservation, transfer, and addition of Gestalt of market data released from social media consumers' living labor and the subsequent sales of outputs for money units. The information release and sales of information is the socially determining factor in the nudgital economy. Price fluctuations may occur through differences in information collected.

The imposition of exclusivity of access to information holds social implications. The possibility of the industrialist-capitalist to use the information and the ability to produce a situation on social media that attracts consumer-workers at a cost to yield an adequate and predictable income or reap their foregone profit holds an implicit social stratification. The hidden possessor's elevated position in the amalgamation of information drives a class division and distinction of authority, which becomes visible in understanding the nudgitalist's actions.

The capitalist social formation includes that the dominant class renews its social control through transforming information into money. Thereby, the end goal of the capitalist-industrialist is to gain as much of the public's information in light of the end goal of accumulation of wealth. In reinventing newly designed information sharing tools and options, there is a systematic encouragement and development of technological capabilities of all kinds to share information and keep the information flow alive. Pleasure of the use value of information sharing yields to an inferior position as capital expands for the information shared and privacy gets infringed upon and people are being vulnerable to be nudged. With the information gathered, consumer profiles can be obtained that help to nudge people into making decisions, may that be purchasing or voting choices. There is also the authority to gather information that may be used against civilians when entering nations at borders or in extending visas or when checking on people's tax honesty. The domination of the nudgital society then lies insofar as there is a right to deny others access based on the interpretation of data that can be used to control the populace and may also be turned against them in democratic votes.

The augmentation of information leads to the manifestation of power of a dominant class of nudgers, who only entertain their workers for the final purpose to generate information that then gets transferred to use value in nudging into purchasing decision and civilian order. Social media is therefore a process that uses online tools to constitute different social classes in their dynamic existence. The owners of social media power are the momentary embodiment of the nudgital society, whereas the users of social media tools are carrying on their activity of production of nudgital information surplus value. Their property of information and privacy gets shared, which allows the amalgamation of information and possession of user profiles and customer and civilian tendencies of the social media capitalist-industrialist, who can then materialize the surplus value gained. The relation between the nudger social media capitalist-industrialist and the nudged is one by hidden domination and exploitation, in which one party holds an amalgam of information about the other that can also be gathered implicitly. That is the power that lies in the nudgital society.

2.1.4 The case of Searchplace Discrimination

Advances in behavioral economics leadership have increasingly created a demand for inclusive digital online searchplaces free from discrimination and harassment. Technology regulation can provide effective relief against negative search engine disoptimization (SEDO). Data regulation adjustment could build on the US Digital Millennium Copyright Act (DMCA) and the EU GDPR (General Data Protection Regulation) and Right-to-be-Forgotten in order to save individual's content being pegged to negative search deoptimization. Standardized content management tools for editing one's own searchplace representation could be coupled with human quality controls. Fast and easy complaint mechanisms are recommended industry changes. Raising awareness can empower consumers to create pressure for industry development.

Contemporary behavioral law and economics developments may aid in envisioning a transition to a more inclusive online searchplace environment. While a legal analysis of online searchplaces grants insights about the disparate impact of online choices, a complementary economic analysis allows to study efficiency of markets with consideration for incentives to curb SEDO and negative externalities of online injustice.

Future behavioral insights may improve societal welfare given novel online inequalities arising from strategic searchplace discrimination. Ethics of inclusion, law and economics scrutiny of digital worlds and interdisciplinary dialogue building but also human artificial intelligence algorithm compatibility, and checks-and-balances are expected to become key advancements in behavioral economics leadership of the future to tackle searchplace discrimination.

2.1.4.1 The rise of online searchplaces

Recent decades have seen breakthroughs in technological developments triggering a digital revolution with palpable consequences in our daily lives. The World Wide Web has become a powerful source of information exchange. By now, the constant access to amalgamated crowd wisdom derived from online searchplaces, such as Google, Bing, Yahoo, DuckDuckGo, Yandex, guides human everyday decisions and choices.

Since the 1990s, online data presentation started to become more systematically structured. Internet search engines were available for the general public that used software systems carrying out searches on the World Wide Web in a consumer-oriented, yet fairly unregulated way.

With exponentially rising information exchange online, human dependency on technological display of information online has risen steadily (The Economist, 2019).[1] The information gain via technological devices turned into the most relevant information source used all the time, often without reflection (Bariffi & Puaschunder, 2021).[2] With the advent of structured data presentation in online search engines, suddenly the entire world began to use online data provision en masse and on a constant basis, imposing a verb to describe this action as "to google" something (Puaschunder, 2022).[3]

With the rise of social online media Internet companies, such as Facebook, Twitter, Instagram, and TikTok, and the popularity of smart devices, we have become familiar and more confident with constant information exchange on public online platforms and public display of personal information. With the advent and massification of so-called social networks—personal portals where people constantly register and consult information from their social environments openly online—search engines to navigate big data on the Internet became the most important information channels in the modern world.

[1] *Slowbalisation: The steam has gone out of globalisation: A new pattern of world commerce is becoming clearer – as are its costs*, THE ECONOMIST (Jan. 24, 2019), https://www.economist.com/leaders/2019/01/24/the-steam-has-gone-out-of-globalisation.

[2] Francisco Bariffi & Julia Margarete Puaschunder, *Artificial Intelligence and Big Data in the age of COVID-19*. RAIS CONF. PROCEEDINGS: 24TH INT'L RAIS CONF. ON SOC. SCI. & HUMANITIES 1, http://rais.education/wp-content/uploads/2021/11/0115.pdf

[3] *Id.*; JULIA MARGARETE PUASCHUNDER, ETHICS OF INCLUSION: THE CASES OF HEALTH, ECONOMICS, EDUCATION, DIGITALIZATION AND THE ENVIRONMENT IN THE POST-COVID-19 ERA (2022).

Search engines boomed in the wake of the COVID-19 pandemic. Systematic lockdowns of entire populations pushed humans to increasingly use digital devices in order to achieve some sort of social connection on a global level. Crowded hospitals and fear of contagion opened gates for e-healthcare and medical forums to gather crowd wisdom on health and well-being advice. Nations with the technological capability used massive surveillance systems of entire populations to control and contain the SARS-CoV-2. Education shifted to remote learning in record speed (Corlatean, 2020).[4]

Today, real-time information collection online holds most valuable insights to guide human interaction, social choices, and everyday behavior on a worldwide scale. The currently breaking 5G technology will enable digitalization to encroach all sorts and tasks of modern life, which is likely to continuously increase the provision and use of information retrieved online for everyday purposes in the future.

2.1.4.2 Searchplaces technicalities

The enormous and constantly rising amount of Internet searchplace activity is supported by artificial intelligence enabling search engine results generation. Search engines are primarily maintained by algorithms and constant web crawlers continuously processing a rising amount of information in real time. The digital revolution generated algorithmic learning technology with processes resembling human decision making, but much faster and often more efficient.

Search engines have grown quantitatively and qualitatively in the last three decades. Computer capacity to process data rose exponentially. As increasingly sophisticated algorithms provided information cheap, quickly, and on a worldwide scale, individuals more and more turned to online search functions as all-around aids in navigating through complex daily lives.

With a rise of information transfer online, search engines became more sophisticated with tools including lists of web pages, images, videos, infographics, articles, research papers, and other types of mined data and files associated with particular search terms. Search results and features differ by search platform but most often include short description snippets, images, maps, definitions, answer boxes, videos, news, blogs, knowledge graphs, discovery, and suggested search refinements or newly emerging vocalized commands.

Since the beginnings of search engines in the 1990s, three features and trends remain at the core of online searchplaces: (1) The relevance of real-time adjustment to keep up with a rising activity online, (2) the commercialization of Internet search functions as markets (Pursel, 2018; Gandal, 2001)[5], and a (3) growing dependence

---

[4] Titus Corlatean, *Risk, discrimination and opportunities for education during the times of COVID-19 pandemic*, RAIS CONF. PROCEEDINGS: 17TH INT'L RAIS CONF. ON SOC. SCI. & HUMANITIES 37, http://rais.education/wp-content/uploads/2020/06/004TC.pdf

[5] Bart Pursel, Search Engines, Penn State Pressbooks (Retrieved February 20, 2018); Neil Gandal (2001), The dynamics of competition in the internet search engine market, 19 IJIO 1103, doi:10.1016/S0167-7187(01)00065-0.

on constantly iterative webcrawlers and algorithms to present an exponentially rising public information exchange on the Internet in a palpable way (Brin & Page, 1998).[6]

The usefulness of search engines depends on the relevance of the results provided to users searching for information.[7] Most search engines employ methods to rank results based on different constantly changing criteria, which are not fully made transparent to the public web audience (Jawadekar, 2011).[8] A mixture of predefined and hierarchically ordered keywords programmed by humans is coupled with an inverted index generated by algorithms that analyze located texts and images.[9]

Competition among search engines has been a driving force in the evolution of search algorithms filtering relevant results with convenience. Searchplaces mainly compete on speed, accuracy, user-friendliness, and specific content-related searches.[10] Search results are partially based on a web of pages, their relevance, and credibility ranking.[11] Usually, searches lead to several pages of descending relevancy and accuracy of contents.

To present search engine results online, mainly automated search engine algorithms rank websites based on a combination of popularity and relevancy (Segev, 2010).[12] Algorithms include factors such as quality and relevance of the content, expertise, authoritativeness, trustworthiness of websites and author on a given topic, good user experience as well as backlinks.[13]

Search engine results pages (SERP) are pages displayed by search engines in response to a query by a user.[14] The results of search engine queries are usually ranked by relevance and credibility. Results usually display titles, links that direct to other pages on the World Wide Web, a short description snippet and descriptions where search terms have matched content with the results page.[15]

---

[6] Brin, Sergey & Page, Larry (1998). The Anatomy of a Large-Scale Hypertextual Web Search Engine, PROCEEDINGS SEVENTH INT'L WWW7 CONF. (Retrieved January 10, 2021); Method for node ranking in a linked database, Google Patents, Archived from the original on 15 October 2015 (Retrieved 19 October 2015); About: RankDex, rankdex.com

[7] Wikipedia entry, Search Engine, Retrieved Aug. 20, 2022, https://en.wikipedia.org/wiki/Search_engine#cite_note-29

[8] WAMAN S JAWADEKAR, KNOWLEDGE MANAGEMENT: TOOLS AND TECHNOLOGY, KNOWLEDGE MANAGEMENT: TEXT & CASES (2011).

[9] Wikipedia entry, Search Engine, *Id.*

[10] *Id.*

[11] *Id.*

[12] EL SEGEV, GOOGLE AND THE DIGITAL DIVIDE: THE BIASES OF ONLINE KNOWLEDGE (2010); Liwen Vaughan & Mike Thelwall (2004), Search engine coverage bias: evidence and possible causes, 40 Info. Process & Mgmt, 693, doi:10.1016/S0306-4573(03)00063-3; Google, How search works (Retrieved Aug. 20, 2022), https://www.google.com/search/howsearchworks/how-search-works/ranking-results/

[13] Google, id.

[14] Wikipedia entry for Search Engine Results Page, id.

[15] *Id.*

2.1.4.3 Searchplace commercialization

Search engine providers generate revenue by commercial ventures—such as advertising, commercialization of listings in search results, search-related advertisements as well as big data analysis.[16] For sponsored results, advertisers choose what to display.

In most recent decades, the placement of homepages and competitive keywords have become an increasingly important field of business and political interest.[17] Sponsored searchplace results are paid prominent information display—mostly on top of search engine results display—sold to the highest bidding entity. Sponsored creative results on Google have become a lucrative business with the most expensive keywords being sold for legal services, especially personal injury lawyers in highly competitive markets.[18] Targeted ads and classified advertisement that picks up searchplace users when searching for specific information have become prominent tools of online guerilla marketing.

The commercialization and the need to generate revenue, however, may conflict with searchplace providers' goal to offer the most useful, healthy and accurate information – for instance when one thinks of advertisement to clog limited human attention span capacity, or the social media paid advertisement scandals. Arguably, commercialization may crowd out the overall integrity and usability of searchplaces for users. In 2018, a European Commission study showed that consumers generally avoid paid top results expecting the sponsoring being less relevant and credible.[19]

2.1.4.4 Systemically biased searchplace discrimination

While digitalization offers unprecedented human advancement in access to worldwide information, recently attention was drawn to systemic biases in searchplaces. Online information display in modern searchplaces has generated notable repercussions for arguments about human biases influencing algorithmic choices.

Empirical studies found political, economic, and social biases in the information search engines provide (Jansen & Rieh, 2010; Segev, 2010; Vaughan & Thelwall, 2004).[20] Technological, political, societal but also economic biases have been addressed as underlying causes (Berkman Center for Internet & Society, 2002; Introna & Nissenbaum, 2000; Jansen & Rieh, 2010; Segev, 2010; Vaughan &

---

[16] How search engine works? (Retrieved June 2018).

[17] Google Warns: Automated Queries On Google Is Against The Terms Of Service, SEROUNDTABLE.COM (Retrieved Apr. 4, 2017), https://www.seroundtable.com/google-warns-automated-queries-23135.html

[18] Wikipedia entry for Search Engine Results Page, *id.*

[19] Behavioural Study on the Transparency of Online Platforms, European Union 19, 2018 (Retrieved Apr. 22, 2021).

[20] SEGEV, GOOGLE AND THE DIGITAL DIVIDE: THE BIASES OF ONLINE KNOWLEDGE (2010); Liwen Vaughan & Mike Thelwall, id.; Bernard J. Jansen & Soo Young Rieh (2010) The Seventeen Theoretical Constructs of Information Searching and Information Retrieval. 61 AM. SOC'Y INT'L L. PROC 1517.

Thelwall, 2004).[21] Biases are likely unnoticingly transmitted by programmers and artificially created big data generated from large-scale preference amalgamation (Bariffi & Puaschunder, 2021; Bartlett, 2018; Roehring & Pring, 2017; Smith & Telang, 2017; Stephens-Davidowitz, 2018).[22]

Online search engine portals have done most important work and attempts toward eradicating or—at least—alleviating biases. At the same time, shifting marketplaces to online virtual spaces opens gates for misinformation and disinformation in search engines and online forums being intentionally used in a competitive, dishonest, and harmful sense.

2.1.4.5 Searchplace discrimination due to search engine exploitation

Besides technically transmitted biases, search engine "rebel" users exploit the structures of search engine algorithms to strategically manipulate search results, often for commercial or strategic reasons.

Search engine optimization (SEO) is the strategic manipulation of search engine results for political, career, social, and commercial advantages. Positive SEO targets at improving the search engine listings of web pages for relevant content search terms.

Negative SEO is targeted at reducing the quality and relevance of search results of professional and commercial competitors. Negative search engine deoptimization (SEDO) refers to strategic searchplace manipulation that aims at changing the perception of a term or push out competition from search engine results (Adams & McCrindle, 2008; Orlowski, 2003; The Register, 2007).[23] SEDO is primarily done for business, political, comical, and competitive purposes (New Oxford American Dictionary; Price, 2005; Search Engine Watch, 2007; The New York Times, 2022; Zeller, 2006).[24]

---

[21] Segev, id.; Liwen & Thelwall, id.; Jansen & Rieh, id.; Berkman Center for Internet & Society (2002), Replacement of Google with Alternative Search Systems in China: Documentation and Screen Shots, Harvard Law School; Lucas Introna & Helen Nissenbaum (2000). Shaping the Web: Why the Politics of Search Engines Matters, 16 INFO. SOC'Y 169, https://doi.org/10.1080/01972240050133634 S2CID 2111039.

[22] Bariffi & Puaschunder, id.; BARTLETT, THE PEOPLE VS. TECH: HOW THE INTERNET IS KILLING DEMOCRACY AND HOW WE SAVE IT (2018); FRANK, ROEHRING & PRING, WHAT TO DO WHEN MACHINES DO EVERYTHING: FIVE WAYS YOUR BUSINESS CAN THRIVE IN AN ECONOMY OF BOTS, AI, AND DATA (2017); SMITH & TELANG, STREAMING, SHARING, STEALING (2017); STEPHENS-DAVIDOWITZ, EVERYBODY LIES: BIG DATA, NEW DATA, AND WHAT THE INTERNET CAN TELL US ABOUT WHO WE REALLY ARE (2018).

[23] Orlowski, Andrew (April 3, 2003). Anti-war slogan coined, repurposed and Googlewashed in 42 days, THE REGISTER. (Retrieved Jan. 6, 2007); ANDREW A. ADAMS & RACHEL MCCRINDLE, PANDORA'S BOX: SOCIAL AND PROFESSIONAL ISSUES OF THE INFORMATION AGE (2008).

[24] Tom Zeller (October 26, 2006). A New Campaign Tactic: Manipulating Google Data, THE NEW YORK TIMES, p.20; Wikipedia entry on Google bombing (Retrieved Aug. 20, 2022), https://en.wikipedia.org/wiki/Google_bombing#cite_note-nytimes1-1; Gary Price (May 16, 2005).

In the case of Google—the most prominent search engine in the Western world—SEDO happens, for example, in the form of Google Bombing or Googlewashing, which causes a website to rank highly in web search engine results for irrelevant, unrelated or off-topic search terms. Spamdexing is the practice of deliberately modifying HTML to place a website close to the beginning of specific search engine results or to assign a page to unrelated content in a misleading or dishonest manner (Gyöngyi & Garcia-Molina, 2005).[25]

While these terms are primarily focused on webpages, individual search terms can also be subject to negative SEDO, which is referred to searchplace discrimination of an individual. Searchplace discrimination is primarily used in political, educational, and academic cases, where reputation capital stakes are high.

So-called search engine deoptimization black hat strategies are competitive and unethical distortions of search engine results that either overemphasize unfavorable search results (likely enabled via clickfarms) or create a misinformation, disinformation, or spam overload that derails from accurate representations of individuals online.

The strategic manipulation of SEO in a harassing, misrepresenting, and discriminatory online display of unassociated, harmful, or violent information is often based on concrete strategies. This hardly discussed and mainly unregulated market information distortion discriminates individuals, who hardly have legal protection, regulatory control mechanisms, or community support from the consumer and individual perspectives.

2.1.4.6 Search Engine Deoptimization (SEDO) technicalities of searchplace discrimination

Searchplace discrimination occurs when individuals are targeted by erasing online quality content information, potentially due to data compartmentalization and wrong flagging of search results as well as overemphasizing negative, unrelated, spamming or harmful contents, potentially via clickfarms and automated bots.

First, negative SEO comes to play in strategically manipulating of breaking big data clouds in combination with pegging harmful, outdated or useless spamming information to actual content keywords describing the victim, so no coherent information is found and no stable knowledge panel is formed. Data compartmentalization may be enabled by capping Google search results at a low number (indicated by Google). The algorithm is tricked to cap the number of search results by infusing critical or unfavored contents (e.g., crimes, hate speech, obscene language, violent, illegal, or sexual content) and then highlighting unfavorable, misleading and/or compromising information via clickfarms in the cropped displayed results. Search engine data cloud compartmentalization may be used to create separate data clouds with positive and negative results, in order to then

Google and Google Bombing Now Included New Oxford American Dictionary, SEARCH ENGINE WATCH. Archived January 27, 2007. Retrieved January 29, 2007.

[25] ZOLTÁN GYÖNGYI & HECTOR GARCIA-MOLINA WEB SPAM TAXONOMY (2005).

overemphasize negative or useless search engine results while shifting quality content into clusters that appear under unrelated keywords.

Second, SEDO can be used to erase online content by inappropriately flagging competitors' quality or favorable contents in order to make competitors' positive image cues disappear and/or the big data of competitors less relevant for search engines. SEDO is possible due to the fact that most real-time searchplaces are enabled mainly based on algorithmic iterative processes and human quality control of flagging is limited. Positive or content information can thus be erased by flagging content online that gets immediately taken offline. The process includes hardly any resuscitation control by human review. While there is a positive possibility to craft the Internet search results together and positive quality control is enabled via flagging online content, reporting inappropriate online content gets blocked immediately. Little quality control is given to this option being used as a strategy to push down or make appropriate content disappear in the wake of competition.

Third, up-playing and aggressively displaying negative contents, calling on hateful online crowds and bad reviews in order to damage competitors' reputations, may be enabled via clickfarms and Amazon mechanical Turk (mTurk), algorithmic repetitive accessing of online contents, and posting in hateful echo chambers of the Internet. Wrong images displayed—potentially brought into Bing searches via related search results of clickfarms in combination with shooting out quality images through backlinks, pegging quality contents to harmful contents in third-party knowledge broker systems (e.g., SlideToDoc, Pagaloo, DATAnalyze, Readcube, etc.) and comment buttons fed with flagged content—are additional black hat techniques to bring down quality search results and elicit a false image in a malicious and/or harmfully competitive way.

These manipulations impose a wide range of damage and negative externalities. For instance, if individual consumer struggles to find credible results. In the case of vulnerable populations, such as children, innocent Internet users may end up traumatized when being exposed to harmful, aggressive, or violent contents. Victims of cyberbullying and searchplace discrimination may face an uphill battle when it comes to critical life decisions determining educational paths and careers. The socio-economic damage may include impairment of careers, scientific advancement but also derailed purchase potential when it comes to searchplace discrimination against promotion and job market candidates, finance professionals, academics and authors or artists selling goods, services, or ideas online.

2.1.5 Digital Inequality in the Post-COVID-19 Era

The legal, economic and regulatory status of digitalization is currently developing in jurisdictions and economies worldwide. The United Nations (UN) agencies and regional organizations descriptively report internationally varying current guidelines, ethics codes, and action statements regarding the digitalization disruption part. The UN is the leading authority on sustainable development, which is targeted by the Sustainable Development Goals, which were incepted in 2015. Strikingly, none of these global goals directly addresses digitalization and the benefits efficient market transitions can hold for economically empowered development. Information and Communication Technology (ICT) is mentioned here or there but rather addressed in a descriptive and sometimes antique way catching up in simply describing the state of the art after the industry development innovation. The United Nations has been criticized by practitioners to have a backwards-looking approach and by regulators to lack any forward-looking market-relevant innovation discourse on digitalization. The United Nations opened a Center on Artificial Intelligence and Robotics within the UN system in the Hague, the Netherlands, in 2017. The International Telecommunication Union worked with more than 25 UN agencies to stage the "AI for Good" Global Summit. The UNESCO has launched a global dialogue on the ethics of AI due to its complexity and impact on society and humanity. The OECD hosted a Council on Artificial Intelligence in the first half of 2019 to set international AI standards on a global level. In 2017 the International Organization for Standardization (ISO) and the International Electrotechnical Commission (IEC) created a joint technical committee to develop IT standards for business and AI consumer applications. Labor unions have also defined critical principles for ethical AI. The United States Library of Congress has comparative e-content and reports on the use of AI in various domains, for instance, health care, currency, and data management.

While all these reports include contemporary accounts of AI, the reports and efforts are different and important but hold a limited view on the role of inequality in digitalization. To this day, no clear account or report exists about the contemporary international approaches to reap the benefits of digitalization and transform societies worldwide through a digitalization revolution. International organizations and global governance focus on digitalization disruption in the 21st century. For instance, none of them addresses the connection between AI and health care moderated by anti-corruption and market financialization. None of the agencies appears to be covering ethics of digitalization. And no account exists on an analysis of the societal downfalls of innovation if access is restricted—besides the access to affordable medicine debate of the 1990s in the World Trade Organization (WTO) discourse and The Agreement on Trade-Related Aspects of Intellectual Property Rights (TRIPS) agreement.

Most recently, the COVID-19 pandemic revolutionized digitalization but also opened eyes to social injustices around the world. Today, the most pressing question of our time is how to use digital innovation to make the world more equitable and align digital efficiency with justice and fairness. No international publication exists to this date about the most novel facets of digital inequality around the world, such as the upcoming 5G revolution, digital warfare, COVID-19 Long Haulers' relief through digitalization but also rights to not be forgotten given algorithmic responses to potentially false competitive claims to remove content from the Internet.

Digital inequality adds a novel angle to economic perspectives of digitalization. To this day, our understanding of the relation of digitalization and inequality is limited. Digitalization is an innovation that constantly evolves. Therefore, holding the most novel angles to where digitalization stands today may naturally add to the existing literature a fresh angle and create a historic landmark account of our contemporary post-COVID digitalization disruption. Besides inequality inherent in big data insights, future digitalization innovations may include advancements in artificial intelligence, 5G, the Internet of things, robotics and supercomputers. Innovation-inherent inequalities are revealed in novel facets of inequality that are particular for digitalization. Besides positive perspective of digitalization innovations, a deeper theoretical system critique highlights the societal implications of inequality. Macroeconomic results but also a more internationally comparative analysis could round up the extensions of our understanding of market innovations' impact on inequality within society.

Empirical macroeconomic research has portrayed unequal access to digitalization and AI as an obstacle to development, for instance in the domain of health care. Artificial intelligence has been found to offer access to education and health care free from anti-corruption and has therefore been praised as a future prerequisite of responsible access to markets, finance and societal welfare [24]. Future research in this area could break the most innovative ground to establish the importance of law and economics for the regulation of a complex and wide-reaching market transition of our lifetimes. An international account of digitalization inequalities with paying close attention to the most contemporary trends arising on the horizon should also cover the role of the COVID-19 market transition for digitalization coupled with the newest innovations and market advancements. The most significant trends in the digitalization revolution feature multiple sectors, such as—for instance—the medical field, cryptocurrencies in the financial world but also access to education and social justice pledges arising worldwide.

Information about the qualitative and quantitative inequality inherent in the contemporary digitalization disruption could serve as a wake-up call for policy-makers and governance leaders but also the wider public to realize wealth inequality around the globe having reached a record high. In a rising wave of inequality research, which appears to have even grown in the aftermath of the COVID-19 pandemic, the novel insights about inequality in the behavioral and macroeconomic domain could stimulate a more qualitative description of inequalities in access to information, efficiency and dignity—all goods that come from or are directly related to digitalization. A better understanding of digital inequalities may also help global governance institutions in providing innovative law & economics-informed public–private partnership solutions for policy-makers.

Connecting all these dots and could serve as a unique historic landmark to capture our contemporary digitalization disruption moment in time but also offer ways to harvest the benefits of innovation ethically and sustainably with respect for inequality aspects. An ample account of the current state of the art of digitalization could prospect future implications or future-oriented policy recommendations. A truly international angle of access to digitalization innovation benefits worldwide offers the first and most needed account of digital social justice as ennobling excellence. An international economic law perspective, which is missing in the market on digitalization and inequality academic research, practitioners' manuals and global governance reports could raise unprecedented attention to the novel connection between digitalization and inequality worldwide. Presenting a range of innovative and unprecedented empirical insights that have direct implications for global governance officials, economics academics, innovation catalysts and policy-makers around the globe would enable deriving leadership and followership implications for using novel technologies daily to improve individual lives, group dynamics, and global governance. International nuances of digital inequality from qualitative and quantitative viewpoints would provide a future-oriented outlook on the digitalization workplace revolution with direct implications for economic growth, international trade and development. First account of the emerging field of Behavioral Law & Economics on digitalization could thereby be fortified with practical examples on the most recent advancements in digitalization commodification. All these endeavors could serve as a guideline for monitoring and evaluation of the current digitalization pegged to social, economic and environmental causes. In its entirety, future digitalization inequality research provides historic landmarks in innovation management of global digitalization reflecting a concerted world effort to improve society in regards to novel and unknown facets of inequality together.

Overall, this book discusses the advantages but also potential downfalls of our contemporary digitalization disruption. Literature and data about digitalization is emerging that outlines that the digitalization disruption is exponentially growing with time, especially in the last decade and that trend has been even accelerated by the global COVID-19 pandemic, which is still not completely over. Within the next five years, the digitalization disruption is likely growing exponentially with 5G being rolled out in major economies of the world. Deepened insights on the societal implications of a wide-reaching digitalization disruption is of value for those interested in innovation but also inequality alleviation. Scholars and practitioners of tomorrow can learn about the socio-economic fallout of digitalization innovation given the market disruption touching on business, law & economics and interdisciplinary practices. The impetus of digitalization rising globally will also entail more and more digitalization encroaching every aspect of human lives. From an international perspective, global audiences ranging from global governance scholars or business executives alike may therefore address the global market disruption from a multidisciplinary angle. Digitalization also underlying today's finance world's potential in a historic market transition will bring increasing and lasting value to historical research on how digitalization shapes modern markets and society.

In all these endeavors, future research may offer unprecedented benefits for international professionals in regard to highlighting the worldwide connection between digitalization and inequality. Backing claims by empirical insights that have direct implications for global health care, economics academics, and policy-makers around the globe may help professionals derive direct leadership and followership implications for integrating digitalization in the modern workspace.

Outlining international nuances of digital inequality but also daring a future outlook on the digitalization workplace revolution, may hold invaluable insights for international trade and global development. When scientific knowledge will be conveyed in tangible and practical examples, the future research may be of value for global professionals and international governance executives alongside academics. Over the course of time, this research can set a historic landmark in our innovation evolution that may hold beneficial insights on management facets of global digitalization. The future research may also aid sustainable development advocates to reflect a concerted world effort to improve novel facets of inequality in strategic stakeholder management. Practitioners may also learn from the first monitoring and evaluation planned of the current digitalization pegged to social, economic, and environmental causes. Timely and historically valuable research may also serve as a trace how important decisions were made during our economic transition time that potentially set the world on a trajectory in regards to the introduction to digitalization into society.

References

1. Armstrong, P., Glyn, A., & Harrison, J. (1991). *Capitalism since 1945*. Basil Blackwell.
2. Bowles, S., Edwards, R., & Roosevelt, F. (2005). *Understanding capitalism: Competition, command, and change*. Oxford University Press.
3. Brenner, R. (2002). *The boom and the bubble: The US in the world economy*. Verso.
4. Brenner, R. (2006). *The economics of global turbulence: The advanced capitalist economies from long boom to long downturn, 1945–2005*. Verso.
5. Camerer, C., Issacharoff, S., Loewenstein, G., O'Donoghue, T., & Rabin, M. (2003). Regulation for conservatives: Behavioral economics and the case for "Asymmetric Paternalism". *University of Pennsylvania Law Review, 151*(3), 1211–1254. https://doi.org/10.2307/3312889
6. Centeno, M. A., Cinlar, E., Cloud, D., Creager, A. N., DiMaggio, P. J., Dixit, A. K., Elga, A. N., et al. (2013). *Global systemic risk*. Unpublished manuscript for research community. Princeton Institute for International and Regional Studies, Princeton University, April 2013.
7. Centeno, M. A., Nag, M., Patterson, T. S., Shaver, A., & Windawi, J. A. (2015). The emergence of global systemic risk. *Annual Review of Sociology, 41*, 1–21. https://doi.org/10.1146/annurev-soc-073014-112317
8. Centeno, M. A., & Tham, A. (2012). *The emergence of risk in the global system*. Princeton University Working Paper.
9. de Rooij, E. A., Green, D. P., & Gerber, A. S. (2009). Field experiments on political behavior and collective action. *Annual Review of Political Science, 12*, 389–395. https://doi.org/10.1146/annurev.polisci.12.060107.154037
10. Evans, A. M., & Krueger, J. I. (2009). The psychology and economics of trust. *Social and Personality Psychology Compass, 3*(6), 1003–1017. https://doi.org/10.1111/j.1751-9004.2009.00232.x
11. Gilpin, R. (2001). Global political economy. *Understanding the international economic order*. Princeton University Press.
12. Held, D., & McGrew, A. G. (2007). *A new world economic order? Global markets and state power: Beyond globalization/anti-globalization: Beyond the great divide*. Polity.
13. Hong, H. G., Li, F. W., & Xu, J. (2019). Climate risk and market efficiency. *Journal of Econometrics, 208*(1), 265–281. https://doi.org/10.1016/j.jeconom.2018.09.015

14. Horberg, E. J., Oveis, C., & Keltner, D. (2011). Emotions as moral amplifiers: An appraisal tendency approach to the influences of distinct emotions upon moral judgments. *Emotion Review, 3,* 237–244. https://doi.org/10.1177/1754073911402384
15. Leonhardt, J. M., Keller, L. R., & Pechmann, C. (2011). Avoiding the risk of responsibility by seeking uncertainty: Responsibility aversion and preference for indirect agency when choosing for others. *Journal of Consumer Psychology, 21*(4), 405–413. https://doi.org/10.1016/j.jcps.2011.01.001
16. Lerner, J. S., Small, D. A., & Loewenstein, G. (2004). Heart strings and purse strings: Carryover effects of emotions on economic transactions. *Psychological Science, 15*(5), 337–341. https://doi.org/10.1111/j.0956-7976.2004.00679.x
17. Marglin, S. (1974). What do bosses do? The origins and functions of hierarchy in capitalist production. *Review of Radical Political Economics, 6,* 60–112. https://doi.org/10.1177/048661347400600206
18. Miller, G., & Rosenfeld, G. (2010). Intellectual hazard: How conceptual biases in complex organizations contributed to the crisis of 2008. *Harvard Journal of Law and Public Policy, 33*(2), 807–840.
19. Minsky, H. P. (1977). Banking and a fragile financial environment. *The Journal of Portfolio Management, 3*(4), 16–22. https://doi.org/10.3905/jpm.1977.408609
20. Moscovici, S. (1988). Notes towards a description of social representations. *Journal of European Social Psychology, 18*(3), 211–250. https://doi.org/10.1002/ejsp.2420180303
21. Okamoto, K. S. (2009). After the bailout: Regulating systemic moral hazard. *UCLA Law Review, 57*(1), 183–236.
22. Paluck, E. L. (2009). What's in a norm? Sources and processes of norm change. *Journal of Personality and Social Psychology, 96,* 594–600. https://doi.org/10.1037/a0014688
23. Puaschunder, J. M. (2015). On the social representations of intergenerational equity. *Oxford Journal of Finance and Risk Perspectives, 4*(4), 78–99.
24. Puaschunder, J. M. (2022). *Ethics of inclusion: The cases of health, economics, education, digitalization and the environment in the post-COVID-19 Era.* Ethics International.
25. Schumpeter, J. A. (1949). *Economic theory and entrepreneurial history.* Harvard University Press.
26. Shaikh, A. M. (2016). *Capitalism: Competition, conflict, and crises.* Oxford University Press. https://doi.org/10.1093/acprof:oso/9780199390632.001.0001
27. Sidanius, J., & Pratto, F. (1999). *Social dominance: An intergroup theory of social hierarchy and oppression.* Cambridge University Press. https://doi.org/10.1017/CBO9781139175043
28. Soskice, D. (1978). Strike waves and wage explosions, 1968–1970: An economic interpretation. In C. Crouch & A. Pizzorno (Eds.), *The resurgence of class conflict in Western Europe since 1968* (pp. 221–245). Palgrave Macmillan.
29. Stiglitz, J. E. (2006). *Making globalization work.* Norton.
30. Summers, L. H., & Pritchett, L. (2012). *Societies of the world: The future of globalization: Issues, actors, and decisions.* Harvard University Faculty of Arts and Sciences.
31. Tajfel, H., & Turner, J. C. (1979). An integrative theory of intergroup conflict. In W. G. Austin & S. Worchel (Eds.), *The social psychology of intergroup relations* (pp. 33–37). Brooks-Cole.
32. Wagner, W., & Hayes, N. (2005). *Everyday discourse and common-sense: The theory of social representation.* Palgrave Macmillan. https://doi.org/10.1007/978-1-137-07263-4
33. Wagner, W., Lorenzi-Cioldi, F., Mankova, I., & Rose, D. (1999). Theory and method of social representations. *Asian Journal of Social Psychology, 2*(1), 95–125. https://doi.org/10.1111/1467-839X.00028

Part III

Behavioral Finance

3 Value at Looking Back

3.1 Reflexivity in Socio-economic Backtesting

Globalization led to an intricate set of interactive relationships between individuals, organizations, and states. Unprecedented global interaction possibilities have made communication more complex than ever before in history as the whole has different properties than the sum of its increasing diversified parts. With growing globalization and quickening of transfer speed but also exacerbated by the digitalization explosion in the wake of the COVID-19 pandemic, information may impose unknown systemic economic risks on a global scale. Collective interaction effects lead to hard-to-foreseeable fallacy of composition downfalls. Emergent risks imbued in interaction appear to be inherent in global economic systems. In light of growing tendencies of globalization, the demand for an in-depth understanding of how information echoes in socio-economic correlates has gained unprecedented momentum. In seeking to shed light on implicit system failures' socio-economic consequences down the road and potentially disastrous outcomes of cumulative actions triggering mass movements, the chapter outlines unexpected dangers and insufficiently described shadows of past market expectation corrections on future economic market performance.

Overall, the following part innovatively paints a novel picture of the mass psychological underpinnings of business cycles based on information flows in order to recommend how certain communication strategies could counterweight and alleviate information failing market performance expectations that could potentially build disastrous financial market mass movements of economic booms and busts. Understanding how information and communication can build socially constructed economic correlates promises to explain how market outcomes are developed in the social compound and can be guided by central agents' communication. Classical theories of price will be reflected in regard to market expectations. Unraveling how central bank economic forecasts produce certain types of price expectations

J. Puaschunder, *Advances in Behavioral Economics and Finance Leadership*, Contributions to Economics, https://doi.org/10.1007/978-3-031-15710-3_3

will allow deriving insights on information forming market patterns, which trigger collectively-shared economic outcomes that may echo in the real economy.

An introduction to the history of economic cycles led to the analysis of the role of information in creating economic booms and busts. After a thorough literature review on the financial market theory with special attention to heterodox viewpoints, the concept of social volatility is introduced and depicted in light of COVID-19. Social volatility adds to quantitative volatility any social aspects that influence and shape economic markets. Integrating social volatility in the application of emergent risk theory onto economic fluctuations offers an innovative way to explain how and what information represented in the media creates economic ups and downs. An introduction to the history of economic cycles includes Soros' [14]. Theory of Reflexivity in order to draw inferences for the analysis of the role of information in creating economic booms and busts in the age of globalization. Empirically, based on a central European central bank's GNP projections and backtesting corrections, a pattern of central bank corrections communication and economic market performance will be unraveled for the first time to outline that central bank market prediction corrections are positively correlated with near-future market performances and negatively correlated with distant future market performances. The collective reality of prices and the irrationality of the crowds perturbating markets will be discussed. Business cycles are argued to obey some kind of natural complexity, as for being influenced by econo-historic communication trends. Recommendations on how to create more stable economic systems by avoiding emergent risks in communicating market prospects more cautiously will be given in the discussion followed by a prospective future research outlook and conclusion.

While by the end of the 1960s, the most renowned economists agreed that recessions were preventable, history proved them wrong [5]. For instance, the period from 1940 to 1973 became renowned for a time of economic prosperity—yet from 1973 a worldwide recession set in. Historic post-World War II booms were transitioned to downturn from the mid-1970s on as the economic performance declined in the industrialized world. While in the 1980s and 1990s, the economy seemed to expand again, from 2000 on productivity slowed, again, and from 1998 to 1999 stock markets and currencies crashed or halted in 2001 and 2008 [14]. Today, supply-side theory explains the downturn dependent on pressures from labor [5]. Behavioral economists give credit to the unplanned, uncoordinated, and competitive nature of capitalist production as well as the problem of aggregates deviating from the individual's choice predictions.

The expectations of outcomes of multiple decisions among people with conflicting interests may not be foreseeable and therefore have been overlooked by stable equilibrium theories. Yet in reality, expectations may guide individual decision-making and therefore impose novel risks in complex markets. Economic indicators of product wages, international competition, output-capital ratio, and post-tax profits have been studied extensively to derive conclusions about economic pre-indicators of crises and recommendations for improving economic systems.

The role of inflation is widely cited in its inverse relation to unemployment to determine economic conditions [2, 4]. International transmissions of inflation are discussed in historic examples to build adaptive expectations on information about current and past experiences with inflation [15]. Disequilibrium inflation is explained in the literature to occur when the actual rate of inflation is greater than the expected followed by social unrest and industrial conflict [15]. Yet in all these discussions, concrete information about economic prospects and the role of economic expectations in shaping ideas and intentions of individual market actors playing out in economic fundamentals is missing.

While business cycle theories primarily focus on describing economic correlates of booms and busts such as tight labor markets, investment trends, and the uneven development throughout the world causing advantages and disadvantages in the economic impact of booms and busts globally [4], less attention is shed on socio-economic correlates that build expectations leading to irrational exuberance.

Communication about markets and their performance is yet hereby argued to be underlying long-term economic trends as well. For instance, information transfer is key for innovation and markets to pick up new ideas. Information shaping expectations are the basis of investment trends. Economic expectations grown out of information on economic prospects and forecasts guide economic market actions. The collective mood in society shapes investment allocations amalgamating into economic trends that determine economic dynamisms of the collective soul of booms and busts. Economic forecasting and information about economic prospects may trickle down in individual economic decision-making, which in its entirety shapes the economy as a whole. Access to information about market performance may also play a key role in price comparisons around the world, which are the basis of outsourcing and capital allocation decisions [4]. Information portrayed in media may also determine the investment mood and credit liquidity preferences of individuals and institutional representatives. Individual communication in the eye of economic ups and downs may in addition be the basis of social unrest and waves of strikes, which have been shown to be underlying factors determining wages and rates of profit for capitalist bleeding into economic cycles [4]. All these correlates set the tact in shaping economic booms and busts as well as the long-term cycles. The role of information for economic long-term cycles, however, has been—so far—widely overlooked in the standard neoclassical literature. Communication interventions are neglected in a wealth of writings on Federal Reserve and Central Bank interventions ranging from lowering interest rates to direct monetary stimulus. Studying the effect of information and communication on economic correlates to implicitly influence investment decisions may offer invaluable insights on how bubbles start and economic fluctuations can be smoothed.

The following chapter therefore departs from the orthodoxy of hyper-rational individuals, who make rational choices based on perfect knowledge purely self-centered not taking other, history, and the governmental-issued information into account. The following empirical study makes a case for past performance and backtesting market prediction corrections influencing future performance. While classical economics typically focuses on actual outcomes rather than on the various expectations that might have motivated them [1], the presented results will connect expectation performance corrections with future market outcomes.

Through capturing the interplay of communication about prospects and the fundamentals of the economy, the following part is meant to shed light on the socio-psychological underpinnings of economic downfalls. The chapter thereby investigates the unprecedentedly described role of information in building and fueling economic booms and downturns. Theoretically, the notion of expectations in classical writings is introduced to then draw on historical foundations in economic analysis as outlined in Soros' [14] *Theory of Reflexivity*. Boom and bust patterns will be theoretically described and be argued to have the expected outcomes deviate from the actual path. The actual path in turn deviates from the underlying fundamentals in reflection of past performances' prediction corrections. The impact of central bank forecasting and backtesting is hypothesized to change the actual performance in shaping economic ups and downs. Empirically, a central bank's economic forecasts of the Gross National Product (GNP) followed by publicly published backtesting corrections being related to certain market outcomes will be portrayed as the instigation of herd and swarm behavior that potentially caused actual economic booms and busts leading to economic crises. Studying the impact of the found future shadow of today's predictions in a heterodox fashion serves as a window of opportunity for alleviating negative externalities of globalization imbued in technocrats' communication about market prospects. Pursuing the greater goal of deriving recommendations on how to stabilize economic markets in the instant communication century will lead to wider recommendations on finding an optimum balance of deregulated market systems and governmental control. The role of social online media's fast-paced communication for creating financial market consumer sentiments will be discussed.

An introduction to the theory of price and history of economic cycles leads to the analysis of the role of information in the creating of economic booms and busts. Economic forecasting and backtesting market correction data of a central European central bank will be presented as for retrieving information on the relation of past performances' corrections and future market outcomes. Recommendations on how to create more stable economic systems by avoiding emergent risks imbued in market communication are given in the discussion followed by a prospective future research outlook and conclusion.

3.2 Results

In most recent decades, a heterodox economics gate has been opened to allow a trenchant real-world relevant analysis of information sharing in financial markets. Contrary to neoclassic economics, this chapter targets at a heterodox economic perspective of business cycles innovatively also shedding light on imbalances based on over- and undershooting expectation corrections. While neoclassical economics begins from a perfectionist base and introduces imperfections as appropriate modifications to the underlying theory, this part highlights the role of information for real competition. Introducing emergent risk mitigation strategies within globalized economic markets may help avert future socio-economic crises. Public trust may be imbued in open market economies through improved economic market stability and societal welfare stemming from universal access to equally-shared benefits of global economies [11].

The following part focuses on representing connections of expectation corrections in backtesting and actual market performance. The chapter thereby targets at explaining the rather unknown consequences of future-oriented market prospect reporting. The neoclassical stylized linear time scale in financial market reportings will be challenged in depicting it to be disrupted by seasonal information shocks. This part shows the problem with the neoclassical assumption of perfect information, which features inconsistent representations of information by shedding light on imperfections that produce certain types of outcomes in consumption, equilibrium, and prices. The book thereby outlines how the market responds to central bank market communications and how market corrections are related to actual market performance in the near future but also backlash to cyclical tendencies in the more distant future. Attention will be paid to temporal heterogeneity, the information blast moment differing from a linear time scale. On a wider scale, the book paints a picture of markets behaving in line with corrected market predictions. Acknowledging that agents make choices under social constraints, a dependence of past performance on current actions will be unraveled. Studying information on market prospects will allow constructing a framework of socially-structured market fundamentals and derive conclusions on how expectation outcomes echo in economics. The importance of historical conditions but also social and cultural structures will be outlined. Thereby, an opening of time consistent predictions will allow contributing to non-linear models of predictions and market outcomes paying tribute to the idea of turbulent real economies. Business cycles will be shown to obey some kind of natural complexity; they are whimsical based on socio-historic and political trends as well as follow the occasional madness of actual human behavior. The monograph thereby embraces diversity in granting heterodox perspectives of our contemporary knowledge on the formation of business cycles. Overall, the book connects microeconomic information flows with macroeconomic fundamentals and addresses the emergent properties of heterogeneous agents through the wide varieties of constructions of expectations.

Departing from classical economics depicting exogenous causes for economic cycles, the following part focuses on unraveling endogenous—thus system-inherent—business cycle drivers. As an alternative to this debate, the following part innovatively paints a novel picture of the mass psychological underpinnings of business cycles in order to recommend certain communication strategies counter-weighting the building of disastrous financial market mass movements. As business cycles are a collective phenomenon, group interactions' potential contribution toward business cycles will innovatively be outlined and the role of information flows among groups uncovered. Studying the role of information in communicating crises appears logic when considering that business cycles are fluctuations found in the aggregate economic activity of nations that organize their work. Information will also set the tact on if and how prices become an abstraction determined by ideology shifts.

The following reflections on the inside of economic ups and downs will introduce psychological elements into economic debates. While economics seem to give clear guidelines on how economic correlates of competition, wage pressure, deregulation, and repression of real wage growth play out in economic terms, the irrational exuberance that leads to purchasing and investment decisions in the overestimation of future profit perspectives cannot be explained by orthodox economics. Booms and busts will be portrayed as phenomena that are built by the collective decision-making within society as market actors anticipate and panic together. The following empirical part will shed light on communication of economic prospects and information representation of estimates of central banks and capture a novel relation of past market predictions correction with near and distant future market performances.

In order to test for the relation of information and expectations on markets shaping prices, information about market projections of a central European central bank was retrieved online.[1] This central bank bi-annually publishes the real Gross National Product (GNP) in percentage changes to the previous year (GNP, *Bruttoinlandsprodukt Veränderung zum Vorjahr in % (real)* in the original) of a central European country for 4 years—1 year retroactively, the year in which the report is issued and the prospect of the following year and the 2-year prospect. The report includes the correction of the former expectation of the GNP in the former calendar year $t-1$, the same calendar year t as the report is issued and the prospect about the future year $t+1$ and 2 years $t-2$ in advance. The information contains the backtesting previous year market performance correction $t-1$, the current year t, the following year $t+1$, and the prospect of market performance in 2 years $t+2$. Information on such market performance and prospects is available since 1998 until now at the central bank's homepage accessible for anyone with access to the Internet. These reports get published twice a year, in June and in December of each year.

The reports offer information about expectations and market prediction corrections. Growth forecast errors have been leveraged into a macroeconomic tool to draw inferences about forecasting model weaknesses in order to make better predictions about future market performances [3]. In the wider sense, forecasting error corrections offer an opportunity to study whether markets move in the direction of announced market corrections.

In the data analysis, at first the market corrections for each data point representing the real Gross National Product (GNP) of a central European national economy in percentage changes to the previous year were calculated. The market corrections were then plotted in relation to the GNP market performance after the correction in the next half year $t+1$ and in the next year $t+2$ period. Graph 3.1 holds the

[1] https://www.oenb.at/Geldpolitik/Konjunktur/prognosen-fuer-oesterreich/gesamtwirtschaftliche-prognose.html.

GNP market performance prediction corrections c and GNP market performance GDP after correction derived from a central European central bank for the years 1999–2017.

The data setup accounted for the corrections c being enacted retroactively in backtesting—so after the time had passed, hence, for example, a correction for 1999 as $t-1$ that was released in t is the year 2000—but the market performance trend being real time, so 2000 market performance trends in 2000 as $t+1$. We can thus infer the relation of a correction for the past performance and the future market performance in the half year $(t+1)$ and the year $(t+2)$ thereafter. Market corrections are represented by variations in estimations over time. Relations between variables are investigated by correlation studies.

A highly significant positive correlation between the overall correction in t and subsequent market performance in the next bi-annual period $t+1$ ($r_{\text{Pearson}(130)} = 0.417$, $p < 0.000$) was found for the years 1999 to 2017. A highly significant negative correlation ($r_{\text{Pearson}(126)} = -0.303$, $p < 0.001$) between the overall correction

Graph 3.1 GNP market performance prediction corrections and GNP market performance after correction (years 1999–2018)

in t and subsequent market performance in the period $t + 2$ ranging from half a year after to a year after the initial correction was found for the years 1999–2017.

In order to determine whether past period $t - 1$ corrections or current state t corrections or future prediction corrections $t + 1$ and $t + 2$ are associated with a certain market trend, correlations were calculated for the association between (1) $t - 1$ corrections and $t + 1$ actual market performance as well as (2) $t - 1$ corrections and $t + 2$ actual market performance (3) t corrections and $t + 1$ actual market performance as well as (4) t corrections and $t + 2$ actual market performance. A significant positive correlation between the correction in $t - 1$ and subsequent market performance in the next bi-annual period $t + 1$ ($r_{\text{Pearson}(37)} = 0.353$, $p<0.032$) was found for the years 1999 to 2017. A highly significant negative correlation between corrections in t and subsequent market performance in the next period starting after half a year up to a year after the announcement $t + 2$ ($r_{\text{Pearson}(37)} = -0.633$, $p < 0.00$) was found for the years 1999–2017.

Overall, retroactive backtesting market corrections are highly significantly positively correlated with the following market performance. Corrections of past periods are positively correlated with current performances in the near future but negatively correlated with future performances in the more distant future. Current period announcements are associated with positive trends in the subsequent period up to half a year after the announcement and negative trends in the period following more than half a year later to a year later. Past market performance shapes the future prospect and expectations about markets.

Growth forecast corrections are calculated by the difference between actual real GDP percentage changes during one period t, based on the latest data, minus the forecast prepared for the period under scrutiny presented in the previous period $t - 1$. $F_{i,t}$ is the forecast error correction in GDP growth for the given periods of t. The associated forecast error correction is $F_{i,t:t} = \{Y_{i,t} | \Omega_t\}$, where $Y_{i,t}$ denotes the market performance in GDP conditional on Ω_t, the information set available in the current t period.

Positives values of $F_{i,t}$ represent an unpredicted outperformance of the market in GDP terms in the next half year. Negative values of $F_{i,t}$ are an unexpected underperformance correction of the market predictions in GDP terms. $F_{i,t:t}$ denotes the forecast error change in GDP growth for the periods t. The standard error term of the regression is $\varepsilon_{i,t}$.

Under the null hypothesis H_0 that forecast expectation corrections have no impact on or relation with actual future market performance, the β-coefficient would be zero.

Data was retrieved from a European central bank's bi-annual GDP growth prospects. A regression to describe the relation of forecast error correction and actual market performance within the next half year ($t + 1$) over 35 data points reveals an overall fit with R-square 0.147 and adjusted R-square 0.096 of the model. The regression coefficient β-value of 0.348 for explaining the market performance in the first period after the announcement of the correction is significant at the 5% one-sided t-testing level $t = 1.148$, with a p-value of 0.039. A one percentage point change in forecasting corrections is associated with a 0.348% point change in GDP growth in the same direction in the first half year following the announcement.

Growth forecast corrections are calculated by the difference between actual real GDP percentage changes during one period t, based on the latest data, minus the forecast prepared for the period under scrutiny presented in the previous period $t - 1$. $F_{i,t}$ is the forecast error correction in GDP growth for the given periods of t. The associated forecast error correction is $F_{i,t:t} = \{Y_{i,t}|\Omega_t\}$, where $Y_{i,t}$ denotes the market performance in GDP conditional on Ω_t, the information set available in the current t period.

Positives values of $F_{i,t}$ represent an unpredicted outperformance of the market in GDP terms in the next half year. Negative values of $F_{i,t}$ are an unexpected underperformance correction of the market predictions in GDP terms. $F_{i,t:t}$ denotes the forecast error change in GDP growth for the periods t. The standard error term of the regression is $\varepsilon_{i,t}$.

Under the null hypothesis H_0 that forecast expectation corrections have no impact on or relation with actual future market performance, the β-coefficient would be zero.

A regression to describe the relation of forecast error correction and actual market performance starting from half a year after the announcement up to a year after the announcement was calculated over 35 data points that reveal an overall fit with R-square 0.414 and adjusted R-square 0.378 of the model. The regression coefficient β-value of -0.648 for explaining the market performance in the second period after the announcement of the correction is significant at the 5% one-sided t-testing level $t = -4.821$, with a p-value of 0.000. A one percentage point change in forecasting corrections is associated with a -0.648% point change in GDP growth in the same direction starting after the first half year after the announcement up to a year after the announcement.

Overall, the history of past predictions and necessary corrections are positively correlated with future market performance for a half year and negatively associated with the period from half a year to a year after the announcement of a market prediction correction.

3.3 Discussion

This chapter serves as an introduction of the idea of an influence of past market expectation corrections on market performance. Past prediction corrections were found to be highly significantly correlated in directionality and strength with aggregate market performance patterns. Past market performance comments were systemically related to aggregate patterns of capitalist economies. The presented results are evidence for the equilibrating process in markets being inherently turbulent. Market outcomes are thereby portrayed as to be more than the sum of their parts, influenced by noise of corrections of past expectations. The interdependence of market actors in prices was shown to be biased by communication creating expectations but also expectation corrections.

The dataset presented is unique insofar as backtesting and reporting of market prediction corrections are not common with some key central banks. For instance, the German Deutsche Bank refrains from backtesting reporting. Yet the results presented offer invaluable insights into the relation of backward corrections of expectations being systemically related to future performance. In the sense of Kahneman and Tversky's (9] prospect theory, past losses loom in future performance.

As for future implications, now that we have outlined that our past predictions' corrections are associated with actual performance, we need to better understand what specific market information can influence market performance. In the future, the concrete use of language as market control may be unraveled. A qualitative study of media contents could enlighten on the concrete and qualitative contents that make a market go up or under. Future prospective research could apply emergent risk theory onto economic fluctuations, which could serve as an innovative way to explain how and what information represented in the media creates economic fluctuations [6]. Linguistic analyses of newspaper articles about the economy could then shed light on how media representations and temporal foci echo in economic correlates and shape market outcomes. As business cycles are a collective phenomenon, group interactions' potential contribution to create business cycles could innovatively be captured in laboratory and field experiments on the role of information flows among groups in creating price expectations. Social discourse forms a social representation. Mapping out the systematic patterns of information flows' impact on economic correlates would allow predictions about processing and classifications of communication and aid in explaining and interpreting economic transformation over time.

Newspapers and online social media accounts are sites and virtual spaces in which the views of various combinations of social forces and practices are articulated. Newspapers directly speak to the groups and organizations to which the readers belong, the institutions, movements, and sections of society they identify. Information thereby shapes social perception. Using linguistic analysis as a way of uncovering the making of economic booms and busts will affect the general consciousness about language as an implicit economic correlate and basic economic fluctuations. This could also lead to a richer understanding of the echo of market reporting on aggregate properties that give rise to stable aggregate patterns. Mass media providing a platform for social discourse to debate ideologies that present information about what is happening, which gives rise to reinterpretation and an expectation in the market should become the study object [8]. Social online media allows anonymous communication, which fosters fast and unhindered international transfer of ideas but is challenged by the partially missing censorship online. To determine the lexical variation and meanings embedded within different linguistic systems of expressing different ideologies or theories will aid to connect the linguistic structure with the social correlates of economic booms and busts. Evaluation of discourse could thereby enlighten on the process of economic formations based on herd mentality and swarm behavior.

Future work may study the linguistic processes to formulate the relation between economic fundamentals and economic outcomes through price. As a configuration of ideas and systems of reality, discourse is a pattern of categorization of complex information. Processes like interpretation, selection, and abstraction are yet shunned from orthodox economic analyses. Further, we do not have information on correspondence as well as linguistic and theoretical processes in the price formation or how linguistic changes can manipulate economic outcomes. Linguistic transformation through focus in time appears important yet is to this day undescribed. The sequence of changes that create booms and busts but also the selection of wording of booms, busts, and crises should therefore become subject to scrutiny. Linguistic changes that determine theoretical and ideological significance could be studied, whereby linguistic discourse was presented as part of an economic analysis. All these endeavors would allow deriving cheap and easily implementable information nudges as countercyclical alleviation of economic frictions.

Since the production of text and the reception in terms of economic correlates are rather unstudied, we need an understanding of how the nature of communicative events influences economic correlates, which may determine differences in economic power and status. Finding the linguistic structures that socialize market reactions will elucidate the role of information in the turbulent construction of the economy. Products of prevailing forms of economic and social organization will help reflect on how social processes and structures are related to material conditions.

The chapter acknowledges the fact that there are different frameworks of interpretation in explaining economic long-term cycles. The societal influences of language on economic correlates should be studied in relation to communication's influence on society's ideological impress. How text embodies interpretations of subjects, evaluations of prices, and relationships between the real economy and the financial community could be described in a qualitative study. As interpretative meanings are created uniquely in time, the systematic use of linguistic structures that are connected with the text's placement should become subject to scrutiny. As in each socio-economic system, there is a social meaning to the natural language and economic communication, which is distinguished in its lexical and syntactic structures articulated. Historical data but also cross-country datasets may reveal what communication facets are particular for a certain economic trend or economic market system. The linguistic structure of the economy and linguistic variations throughout different economic times will help describe how different economic outlooks or circumstances are portrayed and potentially perpetuated by the media.

How individuals' perceptions of the future and the state of the economy influence individuals' spending and investment choices may also have wider implications around the world. A future study on international differences could consolidate a global validity of the results but may also highlight different scenarios around the globe. In some countries, people may end up in an economically unfavorable situation through a self-fulfilling prophecy or self-enforcing mechanism. Developing nations with less fiscal space may face a vicious cycle transmitted through financial markets, where financial stress and macroeconomic self-enforcing feedback mechanisms eliminate the positive impacts of automatic market stabilizers [13]. Contractionary multipliers resulting from a reduction in fiscal spending, which gained attention of EU policy-makers in the aftermath of the 2008/09 World Financial Crisis, may in particular imply negative effects in post-crisis economies [7]. Regime-dependent multipliers weaken economically already left-behind regions even more [10]. How expectations and expectation corrections influence international trade around the globe could be another interesting extension of the first preliminary results.

Future studies may also investigate the temporal foci of communication strategies' impact on current decisions in order to unravel concrete strategies on how market communication should be regulated. Future studies should outline the frequency of specific economic term references and distribution of active or passive, forward-looking or backward-thinking in relation to economic cycles. Since temporal foci were found to play a significant role in tax allocation preferences over time [12], temporal perspectives and temporal bundling strategies for the information sharing of important market prospects should be considered. Neutrality of communication contents should ensure a degree of certainty, continuity, and universality to lower the perturbation biases as well as ideological governance decisions can create. The role of signaling and reputation information for the formation of expectation is another qualitative linguistic research area that could be addressed.

In the wider application, the results mirror the stylized fantasy of market predictions' reality being far from efficient. Communicating expectation corrections gets repackaged at the recipient, whose experience may influence their decision-making and guide actions. Information on corrections may thus shape market outcomes in the sense of reflexivity. For linguistics, the study adds to the meaning and implicit meta-meanings of market prospects on economic correlates which are built upon receipt of messages. Unveiling the reality of communication as driver of economic ups and downs will aid in deriving communication recommendations to ease economic fluctuations. Central planners but also journalists should be enabled to understand the economic ethos of words and the moral imperative of their economic coverage.

In this sense, access to market information determines the distribution of power within society. Central banker's privileged access to information but also governmental insights, technocrats' knowledge, and media moguls control through language draw a hierarchy of those who hold information before it reaches the general populace. The implicit meaning of the results triggers us to re-evaluate economic systems for a democratization of information and derive concrete communication strategy recommendations. Central bankers could embrace a culture of an information fiduciary to responsibly communicate market information and control the potential fallout pro-actively. Leakage of sensitive information and fraud through insider trading based on information may be an essential boundary condition that deserves closer scrutiny in future work in this domain. Fake news and deep fake misleading information's emergent risk potential should become integrated into macroeconomic frameworks and legislative control should be adapted to the potential threat of new social media tools creating expectations. Due diligence of information provision and accuracy should become part of corporate governance frameworks. Artificial intelligence ethics should embrace information quality control mechanisms in Fintech solutions. Nations around the world may imbue in the hallmarks of democracy and free markets the democratization of information meaning a fair and free access to accurate information to all presented in a cautious, correctly informed, and forward-thinking way.

In the history of globalization, unprecedented global interaction possibilities have made communication more complex than ever before right now as the whole has different properties than the sum of its increasing diversified parts. While deglobalization and slowbalisation trends have become noticed since 2010 in classic globalization hallmarks such as international trade, cross-border foreign direct investments, and global value chains; data transfer and the reaping of surplus value from big data analyses and targeted online advertisement skyrocketed in the last decade. With growing digitalization and quickening of transfer speed, information may impose unknown systemic economic risks on a global scale. In a socially distanced COVID-19 world, we have become virtually closer and digitally more connected than ever before. But today's digital collective interaction effects may at the same time lead to hard-to-foreseeable fallacy of composition downfalls. Emergent risks in emotional communication reactions and interactions appear to imbue inherent social volatility into global economic systems.

In light of growing tendencies of data transfer during globalization, the demand for an in-depth understanding of how information echoes in socio-economic correlates has gained unprecedented momentum. Collective moods, and in particular anxiety, fear, and Angst but also euphoria and hysteresis may play a fundamental role underlying economic cycles, which have hardly been described or empirically validated before. Shedding light on communication causing implicit system failures helps understanding potential socio-economic consequences. Uncovering unexpected dangers and insufficiently described shadows of the invisible hand saves from potentially disastrous outcomes of cumulative moods triggering mass movements that cause economic turmoil.

Paying attention to the currently ongoing novel Coronavirus pandemic may thereby serve to capture our contemporary Zeitgeist. Special attention paid to communication during times of crises and information transfer's impetus in the age of digitalization can innovatively paint a novel picture of the mass psychological underpinnings of business cycles based on information flows with particular attention to digital communication. Addressing problems of the neoclassical assumption of perfect information markets through the lens of "real competition" will specifically unravel how contemporary media communication produces certain types of expectations that form collective moods that then may change consumption and economic interaction patterns resulting in systemic global economic outcomes.

In light of growing tendencies of data transfer during globalization, the demand for an in-depth understanding of how information echoes in socio-economic correlates has gained unprecedented momentum. Collective moods, and in particular anxiety, fear and Angst but also euphoria and hysteresis may play a fundamental role underlying economic cycles, which have hardly been described or empirically validated before. Shedding light on communication causing implicit system failures helps understanding potential socio-economic consequences. Uncovering unexpected dangers and insufficiently described shadows of the invisible hand saves from potentially disastrous outcomes of cumulative moods triggering mass movements that cause economic turmoil.

Paying attention to the currently ongoing novel Coronavirus pandemic may thereby serve to capture our contemporary Zeitgeist. Special attention paid to communication during times of crises and information transfer's impetus in the age of digitalization can innovatively paint a novel picture of the mass psychological underpinnings of business cycles based on information flows with particular attention to digital communication. Addressing problems of the neoclassical assumption of perfect information markets through the lens of "real competition" will specifically unravel how contemporary media communication produces certain types of expectations that form collective moods that then may change consumption and economic interaction patterns resulting in systemic global economic outcomes.

Empirically, the current COVID-19 pandemic serves as an example of an external shock coming down on society with direct impact on societal moods and subsequently connected economic changes. Future qualitative case studies may scrutinize different country regimes' communication strategies in order to underline and nuance the various approaches taken to introduce the external shock of a pandemic all over the world. Empirically, linguistic analyses of news reports about the societal situation but also the economy could be aimed at shedding light on how media representations echo in economic correlates and shape market outcomes. Business cycles could be investigated to obey some kind of natural complexity, as for being influenced by socio-historic trends, political strategies, and specific media communication. Highlighting country differences and regime-specific facets of novel communication techniques will elucidate that varieties of digitalization exist over the world. Paying special attention to different uses of the Internet and multiple communication tools around the global will help to infer about classified contents' influence on economic outcomes. Categorizing communication contents in qualitative analyses based on temporal foci and emotionality will offer to detect positive media strategies and communication contents that allow to stabilize economies in case of an external shock.

In addition, a quantitative cross-sectional study could compare country differences in COVID-19 pandemic communication strategies' impact on economic performance while controlling for the lockdown strategy and infected population per inhabitant as moderator variables. As business cycles are a collective phenomenon, group interactions' potential contribution to create business cycles could innovatively be outlined and the role of information flows among groups in creating price expectations and subsequent market performance unraveled. Special attention should be paid to the degree of digitalization prevalent in different territories of the world in order to find out the impact of digitalization on economic market performance. The analysis would also reveal if instant communication implies either crowd control security to stabilize markets or collective mood variance in causing emotional frictions and implying social volatility causing economic turmoil.

Future behavioral macroeconomic approaches could feature a time window technique to calculate the similarity of time window episodes based on standard economic correlates that capture economic fluctuations. The first time window t_1 could range from 2006 to 2011 to snapshot the 2008/09 World Financial Recession economic fallout. Similarity to a second time window t_2 that includes the period from 2017 until shortly before the outbreak of the COVID-19 pandemic (another time window t_2 ranging from 2017 to 2019) could be measured with Pearson's correlations for each time sequence of measurement interval within the windows. Backtesting of market predictions and economic time series extrapolations would reveal the inherently instable character of markets featuring boom and bust fluctuations. A third time window t_3 could capture the period from the outbreak of the pandemic in order to draw inferences about potential recovery strategies and find either similarities or substantial differences with the 2008/09 World Financial Recession recovery period through using Pearson's correlations by overlaying the two windows t_1 and t_3 for each time sequence of measurement interval within the windows. Any potential differences between t_1 and t_3 will be explained by the contemporary role of digitalization but also found in the differences between the capitalism-inherent crisis captured in t_1 and the currently exogenous shock recovery t_3 period. The results of comparing the capitalism-inherent crisis of 2008/09 World Financial Recession recovery with the external-shock pandemic economy market recovery will serve as invaluable opportunity to differentiate between two timely close (almost all else held equal condition) but systemically differing (internal capitalism failure vs. external shock fallout) cause of economic downfall. The aspired results will offer implications for the nature of capitalism but also add to endogenous and exogenous growth and economic cycle theories.

Overall, future research could acknowledge that human beings' connection and interaction with each other in families, ties and larger networks of states, nations and intergovernmental institutions results in socially constructed volatility that echoes in economic correlates. Understanding how the social compound forms economic outcomes promises to explain how market outcomes are developed in society and can be shaped by strategic communication with special attention to new media technologies. Future research discussions could highlight the uniqueness and differences of the 2019 COVID pandemic economic fallout from the 2008/09 World Financial Recession. While the previous recession was inherent in the system of capitalism and largely played out in financial constraints affecting almost all industries, the novel 2019 COVID pandemic imposes an external shock that was fueled by communication and interaction, which differed throughout the world. The interpretation of the pandemic serves as historic trace in time.

The conservation of the currently ongoing market communication offers important insights for building the fundamental architecture of future more stable markets to be crafted by heterodox economists and off-mainstream public policy experts.

External market shock analyses during the COVID-19 crisis should also target at highlighting winning and losing industries to follow the mandate to find creative strategies how to redistribute the gains of the crisis in order to offset the economic fallout losses in proposed inequality alleviation strategies. This kind of research should also discuss the contemporary divide between economic fundamentals, financial market performance versus the real economy fallout. All these endeavors target at deriving how certain communication strategies could counterweight and alleviate the building of collective moods bleeding into disastrous mass movements causing turmoil in financial market and steering economic fallouts with negative implications for societies' weakest segments. A prospective future research outlook and implications should be offered for the economic future of health care, workforce, city planning, and education in our post-COVID-19 world to come based on recommendations how to build stable economic systems by avoiding emergent risks and communicating market prospects favorably.

References

1. Arestis, P., & Eatwell, J. (2008). *Issues in finance and industry: Essays in honour of Ajit Singh.* Palgrave Macmillan.
2. Armstrong, P., Glyn, A., & Harrison, J. (1991). *Capitalism since 1945.* Basil Blackwell.
3. Blanchard, O., & Leigh, D. (2013). *Growth and forecast errors and fiscal multipliers.* IMF Working Paper, Washington, DC.
4. Brenner, R. (2002). *The boom and the bubble: The US in the world economy.* Verso.
5. Brenner, R. (2006). *The economics of global turbulence: The advanced capitalist economies from long boom to long downturn 1945–2005.* Verso.
6. Centeno, M. A., Creager, A. N., Elga, A., Felton, E., Katz, St. N., Massey, W. A., & Shapiro, J. N. (2013). *Global systemic risk: Proposal for a research community.* Unpublished working paper, Princeton University Institute for International and Regional Studies.
7. European Commission. (2014). *Financial crisis: Causes, policy responses, future challenges: Outcomes of EU-funded research.* EUR 26554 EN, Research and Innovation.
8. Fowler, R., Hodge, B., Kress, G., & Trew, T. (1979). *Language and control.* Routledge.
9. Kahneman, D., & Tversky, A. (1979). Prospect theory: An analysis of decision under risk. *Econometrica, 47*(2), 263–291. https://doi.org/10.2307/1914185
10. Mittnik, S., & Semmler, W. (2012). Regime dependence of the multiplier. *Journal of Economic Behavior & Organization, 83*(3), 502–522. https://doi.org/10.1016/j.jebo.2012.02.005
11. Puaschunder, J. M. (2015). Trust and reciprocity drive common goods allocation norms. In *Proceedings of the Cambridge business & economics conference,* Cambridge University. Proceedings of the 2015 6th international conference of the Association of Global Management Studies at Alfred Lerner Hall of Columbia University, New York. The Association of Global Management Studies. Oxford Journal: An International Journal of Business & Economics.
12. Puaschunder, J. M., & Schwarz, G. (2012). *The future is now: How joint decision-making curbs hyperbolic discounting but blurs social responsibility in the intergenerational equity public policy domain.* Harvard University Situationist Law and Mind Sciences Working Paper.
13. Semmler, W. (2013). The macroeconomics of austerity in the European Union. *Social Research, 80*(3), 883–914.

14. Soros, G. (1994). *The theory of reflexivity*. The MIT Department of Economics World Economy Laboratory Conference, Washington, DC. Retrieved at http://mertsahinoglu.com/research/the-theory-of-reflexivity-by-george-soros/
15. Soskice, D. (1978). Strike waves and wage explosions, 1968–1970: An economic interpretation. In C. Crouch & A. Pizzorno (Eds.), *The resurgence of class conflict in Western Europe since 1968* (pp. 221–245). Palgrave Macmillan.

Financial Behavioralism: A Behavioral Finance Approach to Minimize Losses and Maximize Profits from Heuristics and Biases

4

4.1 Diversifying Nudges

When going on vacation and not knowing what the weather will be like, you better pack sunscreen and an umbrella. Diversification uses the same rational. When not knowing and unable to be influencing the overall market performance one should be prepared for both—market upswings and downturns. Diversification is a risk management technique to mix a variety of preferably contrary investments within a portfolio. A diversified portfolio featuring different kinds of investments will, on average, yield higher returns and pose lower risks than any narrow, single individual investment [57]. Diversification thereby smooths out unsystematic risk. The different contrary options within a portfolio neutralize each other. Benefits of diversification hold if securities in one portfolio are not perfectly correlated, e.g., investing in domestic and foreign markets at the same time or betting on upswing and downswing options of markets concurrently.

Mutual funds are an easy and inexpensive source of outsourced diversification that has gained popularity after the 2008/09 World Financial Crisis. While mutual funds provide diversification across various asset classes, exchange-traded funds (ETF) afford investor access to narrow markets such as commodities and international plays that would ordinarily be difficult to access. To ensure true diversification, divergent correlations among securities have to be achieved. What can we learn from diversification for nudging people into better choices?

Diversification was also key during the Coronavirus crisis is an external shock coming down on society with direct impact on societal moods and subsequently connected economic changes. In contrast to earlier system-inherent economic turmoil resulting in financial sector induced liquidity constraints, the external COVID shock caused "social volatility"—a collectively depressed mood that largely dampened consumption. The difference to previous systemic recessions can be seen in the rapid recovery of well-managed financial funds—for example, the S&P 500

J. Puaschunder, *Advances in Behavioral Economics and Finance Leadership*, Contributions to Economics, https://doi.org/10.1007/978-3-031-15710-3_4

recovered 50% of its pre-COVID value within the first three months after the crisis and reached an all-time high in August 2020. Deutsche Bank recorded rising earnings during the ongoing Coronavirus crisis, with its investment bank branch leading with 43% or 2.4 billion euros revenue [24].

The clear distinction between COVID-19 profit and loss industries made it possible for today's highly flexible financial world to exchange underperforming market segments—such as oil, public transport, and aviation, face-to-face service sectors such as international hospitality and gastronomy—with outperforming market options—such as pharmaceuticals and emergency devices for health care, digital technologies, fintech, artificial intelligence, and big data analytics industries, online retail, automotive and interior design and architecture. COVID-19 now not only created significant health and security risks, social discrimination, and economic costs, but also brought about unanticipated opportunities. Industries profiting economically from the pandemic are comprised of hygiene, pharmaceuticals, and the medical professions [1, 54]. Based on the COVID-19 economic fallout, the finance world had opportunities to diversify portfolios and exchange losing industries for winning market options. That way, a long-term economic downturn could be averted as the financial market jumped up and regained almost 50% of its pre-COVID-19 level within the first three months after the initial downturn in the eye of a global pandemic. In contrast, the real economy was more inflexible with long-term contractual agreements and industrial ties.

In the real economy, concrete constraints create a more emotional and destructive reaction to the general information about COVID-19. Social online media plays a role in loading these two groups. Comparing the economic consequence of the endogenous crunch of the 2008 World Financial Recession with the external economic shock of the COVID-19 pandemic aids to retrieve crisis-specific recovery recommendations in the overall discussion. Understanding how the social compound forms economic outcomes promises to explain how market outcomes are developed in society and can be shaped by strategic communication with special attention to new media technologies on the long run.

Intertemporal choice structures have shown that when individuals judge alternative choices, their decision-making is prone to be biased when evaluating alternatives one at a time. Contrary to standard utility theory would predict, presenting joint alternatives concurrently changes decision-making outcomes toward people becoming more likely to make more rational choices [32, 122]. A natural tendency toward evaluating choices jointly rather than separately improves the quality of decisions as it alleviates complexity and allows to trade-off alternatives directly [44, 46]. Decision-making failures can thus be curbed when several choices are presented together. Intertemporal discounting and intergenerational equity research find that human capacities to consider future outcomes in today's decision-making are limited [51, 95, 96].

The hyperbolic discounting literature describes human decision-making to be constrained over time [51, 59] and shows that people tend to choose patiently when deciding for the future and impatiently when choosing for the present. Field and laboratory experiments provide widespread empirical evidence for this discounting bias ranging from savings [17, 52, 117], credit card borrowing [62, 107], and financial investment. Unraveling ways how to improve impulsive decision-making and nudge people into foresighted control promises a cost-effective means to better day-to-day decisions. Different interventions have been proposed to curb harmful impulsivity. Long-term visions and future-oriented planning changed by commitment, goal setting, planning, and incentives promise to improve decisions [5, 8, 31, 43, 47, 121].

Behavioral financiers knowing about the joint decision-making advantage may choose their financial allocation options together. Joint decision-making advantages may thus become a powerful means to overcome narrow investment options. Bundling of two alternate outcomes of the own performance has also proven to offset separate bills' costs while preserving their net benefits. Bundled legislations were favored over their individual components and increased the psychological willingness to accept alternative perspectives and to embrace diverse stakeholders' viewpoints. In intertemporal predicaments, presenting two temporal snapshots concurrently could serve as a means to overcome intertemporal decision-making biases [93]. A concurrent presentation of options may thus lead to a more diversified portfolio choice.

When facing intertemporal dilemmas, presenting two generational points of view concurrently could be a useful tool to make the future present during the time of the decision. Concretely in financial predicaments, eliciting different decision outcomes concurrently could implicitly lead decision-makers to make less intertemporally-biased choices. For instance, financial decision-makers could envision that they save money according to their personal plans and as anticipated. Well-informed behavioral financiers, however, could also prospect that they will fail to stick to their plan. Considering these two scenarios together will likely help curbing unfavorable overconfidence and aid decision-makers making wiser choices.

Joint decision-making could thereby serve as a means to overcome intertemporal decision-making biases and in particular help implementing real-life relevant financing strategies. Extending the intertemporal choice literature and bundling strategies, future research could examine how presenting positive and negative outcomes in the wish to stick to self-imposed financial plans in order to show that this strategy helps gain a more widespread decision perspective that saves people from harmful myopic decisions. The temporal policy bundling strategy could aid as a powerful tool for implementing disciplined choices in the finance domain.

4.2 Crises-Robust Market Options

Today, social responsibility has emerged into an *en vogue* topic for the corporate world and the finance sector. Contrary to classic finance theory that attributes investments to be primarily based on expected utility and volatility, the consideration of social responsibility in financial investment decisions has gained unprecedented momentum (The Economist, January 17, 2008; The Wall Street Journal, August 21, 2008).

Financial social responsibility is foremost addressed in Socially Responsible Investment (SRI), which imbues personal values and social concerns into financial investments [75, 84, 85, 101]. SRI thereby merges the concerns of a broad variety of stakeholders with shareholder interests [113]. SRI is an asset allocation style, by which securities are not only selected on the basis of profit return and risk probabilities, but foremost in regard to social and environmental contributions of the issuing entities [6, 124]. SRI assets combine social, environmental, and financial aspects in investment options [23, 36].

Socially responsible firms receive more lenient settlements from prosecutors and have higher resulting market valuations [42]. A one standard deviation increase in CSR is associated with 5 million dollars less in fines, or 25% lower than the mean and less costly subsequent monitoring. High CSR firms outperform low CSR firms by 2.4% in the 6 months following the announcement of the settlement [41, 42].

The consideration of Corporate Social Responsibility in investment decisions is the basis for Socially Responsible Investment (SRI). SRI is an asset allocation style, in which securities are not only selected for their expected yield and volatility, but foremost for social, environmental, and institutional aspects [69, 84, 85]. The most common forms to align financial investments with ethical, moral, and social facets are socially responsible screenings, shareholder advocacy, community investing and social venture capital funding. SRI is a multi-stakeholder phenomenon that comprises economic, organizational, and societal constituents. In recent decades, SRI experienced a qualitative and quantitative growth in the Western World that can be traced back to a combination of historical incidents, legislative compulsion, and stakeholder pressure. SRI is a context and culture-dependent phenomenon that seems to stem out of personal ethics and values that supplement profit maximization goals [69].

Socially responsible screenings are "double bottom line analyses" of corporate economic performance and social responsibility. In screenings, financial market options are evaluated based on economic fundamentals as well as social features and corporate conduct externalities [101]. In addition to the traditional scanning of expected utility and volatility, screenings include qualitative examinations of intra- (e.g., corporate policies and practices, employee relations) and extra-organizational (e.g., externalities on current and future constituents) features of corporate conduct [101]. In general, screenings are based on corporate track records of societal impacts, environmental performance, human rights attribution, and fair workplace policies as well as health and safety standards outlined in CSR reports. Consequentially screening leads to the in- or exclusion of corporations from portfolios based on social, environmental, and political criteria. Positive screenings feature the selection of corporations with sound social and environmental records and socially responsible corporate governance [97]. Areas of positive corporate conduct are human rights, the environment, health, safety, and labor standards as well as customer and stakeholder relations. Corporations that pass positive screenings meet value requirements expressed in their social standards, environmental policies, labor relations, and community-related corporate governance [71]. Negative screenings exclude corporations that engage in morally, ethically, and socially irresponsible activities. Pro-active negative screenings refrain from entities with corporate conduct counterparting from international legal standards and/or implying negative social externalities [97]. Negative screenings may address addictive products (e.g., liquor, tobacco, gambling), defense (e.g., weapons, firearms), environmentally hazardous production (e.g., pollution, nuclear power production), but also social, political, and humanitarian deficiencies (e.g., minority discrimination, human rights violations). Specialty screens feature extraordinary executive compensations, abortion, birth control, animal testing, and international labor standard infringements [23]. Post hoc negative screening implies divestiture as the removal of investment capital from corporations and/or markets. Divestiture is common to steer change in politically incorrect regimes, but also used to promote environmental protection, human rights, working conditions, animal protection, safety, and health standards [13, 36, 60]. Political divestiture describes foreign investment flight from politically incorrect markets based on CSR information [113]. Political divestiture targets at forcing political change by imposing financial constraints onto politically incorrect regimes that counterpart from international law resulting in war, social conflict, terrorism, and human rights violations [69, 71, 75, 76, 90]. Divestiture cases are South Africa during the Apartheid regime; governmental human rights violations in Burma as well as the current financial movement from fossil fuels[70, 72].

Positively screened SRI funds are more likely to feature IT-technology and alternative energy industries that attract innovative venture capital providers. Positively screened SRI options tend to be more volatile, yet if successful, grant high profitability—e.g., solar energy funds have significantly outperformed the market in recent years and remained relatively stable during the 2008 financial crisis. As for excluding high-return, high-volatility industries such as petroleum, defense, and addictive substances, negatively-screened options are more likely to underperform the market, and at the same time are robust to overall market changes. Negative screening asset holders are more loyal to their choice in times of crises, which contributes to the stability of these funds. Data on the profitability of political divestiture indicates a potential first-mover advantage for early divestiture [69, 74, 75, 78].

Behavioral finance strategists consider the ethical roots of SRI. SRI can be traced back to ethical investing of religious institutions and societal attention to social, environmental, and political deficiencies [70]. In the 1960s, shareholder activism of civil rights campaigns and social justice movements drove SRI. Since the 1980s, positive screenings identified corporations with respective CSR policies and political divestiture became prominent in the case of South Africa's Apartheid regime [78]. Environmental catastrophes in Chernobyl and Bhopal as well as the Exxon Valdez oil spill triggered environmentally conscientious investment. SRI was propelled in the wake of the microfinance and cooperative banking revolution. To this day, SRI is connected to global governance, for instance, in the United

Nations having launched "The Principles for Responsible Investment" in collaboration with the NYSE in 2006. In the wake of the 2008 financial crisis, SRI is attributed to the potential to re-establish trust through stability in financial markets [82].

SRI motives are proposed comprising—apart from profitability calculus—socio-psychological motivating factors such as altruism, innovation and entrepreneurship, strategic leadership advantages, information disclosure, self-enhancement, and expression of social values of socially responsible investors, who have a long-term focus [76, 77, 82, 88, 89). SRI options fulfill a need for transparency and information disclosure and are therefore strategies to diminish uncertainty in purchase decisions.

In a cost and benefit analysis, SRI implies short-term expenditures but grants long-term sustainable investment streams. In the short run, screened funds have a higher expense ratio in comparison to unscreened ones—that is, social responsibility imposes an instantaneous "ethical penalty" of decreased immediate shareholder revenue [63, 119]. In addition, for investors, the search for information and learning about CSR is associated with cognitive costs. Screening requires an extra analytical step in decision-making, whereby positive screens are believed to be more cognitively intensive than negative ones [55]. Screening out financial options lowers the degrees of freedom of a full-choice market spectrum and risk diversification possibilities [9, 63, 124]. On the long run, SRI options offer higher stability, lower turnover, and failure rates and litigation or consumer boycott risks compared to general assets [22, 28, 33, 100, 115]. Being based on more elaborate decision-making processes, once investors have made their socially responsible choice, they are more likely to stay with it [55]. As a matter of fact, SRI options are less volatile and more robust during cyclical changes [11].

Behavioral finance specialists could not only combine and diversify the mentioned options for positive and negative screening. They could also seek to evaluate options concurrently to make the better choice.

The basis for shareholder activism is transparency and information disclosure, monitoring of corporate conduct, accountability of the implementation of corporate codes of social conduct as well as internal and external CSR monitoring systems. Especially in the wake of the 2008 financial crisis, corporate governance failures and responsibility deficiencies of market actors have pushed investor calls for transparency of corporate conduct, accountability of shareholder meetings, standardized tracking of proxy votings, and accessibility of shareholder meetings. Access to information is believed to lower economic default risks of socially irresponsible corporate conduct and contribute to SRI trends. Financial market disclosure regulations were installed to prevent future economic turmoil due to financial fraud and principal-agent defaults. As a positive externality of the 2008 financial crisis, the drive toward transparency and accountability within financial markets is likely to foster SRI in the future.

Financial social responsibility also allows investors to attribute causes that are in line with their beliefs and societal values. SRI combines financial investments with personal values based on societal ethicality [2, 27, 111]. As a means to integrate ethicality in economic decision-making, SRI enables investors to address protected ethicality notions that are in line with their personally-held, culturally-established social values [50].

Investment decision-making depends on information about corporate conduct. Information on corporate social conduct is a prerequisite for investors' trust in corporations, lowered stakeholder pressure, and litigation risks. Information on CSR impacts on investors' behavior and triggers financial social responsibility [30, 64, 65, 83, 108, 124]. Investors' access to information about CSR is a prerequisite for SRI. SRI is based on disclosure of corporate social conduct [20, 55, 64). In general, consumers' knowledge about the CSR performance heightens the positive perception of corporations and triggers investment endeavors.

The latest trends after the COVID-19 pandemic have stressed attention to social inequality [92]. In the course of the crisis, some industries have been able to increase their market share and profits in the wake of widespread consumption changes—for instance, private medical care and pharmacy, hygiene, digitization and ecological nutrition are winning industries during COVID-19. At the same time, some industries have collapsed due to the external shock of the coronavirus pandemic and the associated restrictions such as lockdowns, foremost gastronomy, air traffic, tourism and event management [91].

This unequally distributed burden of the COVID pandemic implies that financial markets have been able to recover from the crisis relatively quickly. For example, the Standard & Poor's (S&P) 500 Index has recovered by around half of its previously-fallen value just three months after the markets slumped in February and March 2020 and has recorded market high trends since summer 2020.

Financial market actors can easily diversify options and swap COVID-losing options for winning ones in index funds. Financial market agents can also speculate on a falling value of industries and thus turn an anticipated fall in prices into a profit. At the same time, a gap has opened up to the real economy, which is often confronted with unemployment or short-time work during the crisis. This imbalance between the financial world and the real economy in the course of the COVID-19 crisis is also exacerbated by a lower key interest rate and high inflation as a result of the economic aid packages provided worldwide.

Future advancements of the idea of CSR and SRI are enacted in the United States Green New Deal (GND) and the European Union European Green Deal with the European Finance Taxonomy. The United States serves as market solution to implement global environmental governance as "the sum of the many ways individuals and institutions, public and private, manage their common affairs." Based on historical foundations, the underlying economic mechanisms of the GND and contemporary implementation strategies of the GND are John Maynard Keynes' [48] multiplier effect, which speaks about governmental investments trickling down through all layers of the economy, and the Sustainable Development Goals.

The downside of the governmental rescue and recovery aid is the triggering of a long-term low interest rate in order to incentivize investment capital on financial markets. At the same time, however, households are incentivized to go into debt through cheap lending in the real world in connection with inflation reducing their savings in the bank account. This situation makes real-world households more vulnerable and less resilient in future unforeseeable shock situations.

This in turn increases the probability of behavioral decision-making fallibilities in the real world, such as taking out too many loans, which can then lead to payment difficulties in the long run and also imply stress levels' pushing propensities to unhealthy nutrition and drug consumption in vulnerable populations [7, 14]. The unequal effect of the pandemic on different social groups underlines the need for a multi-layered analysis of the influence of economic conditions on different social strata.

Corporate social justice mandates and promotion of inclusive excellence with regard to the social, economic and societal added value of diversity are the trends of our time that have emerged from this situation [92, 126]. Above all, the value and the decisive role of educational institutions is emphasized in manifesting and sustainably promoting excellence in diversity, but also in living inclusion within society holistically [92].

4.3 Long-Term Sustainable Market Options

As for being incentivized by first-mover leadership advantages, more and more corporations may pay attention to social responsibility and social inclusion as well as diversity in the future. Accompanied by followers, the rising supply of SRI in combination with a heightened demand the integration of personal values and societal concerns into financial decision-making may prospectively leverage social conscientiousness to become a standard feature of investment markets. On the long run, the integration of SRI into the overall competitive model will further sophisticate social responsibility and diversity in corporate conduct [101, 112, 114, 126]. Financial market demand and supply geared toward SRI and inclusive social justice will stretch the option range in a more socially responsible direction. In addition, if the majority of investors are socially conscientious, socially responsible corporations will continuously benefit from increasing investment streams. Diversity market gains through additional diversification potential will prospectively lead to broad-based acceptance of corporate social justice activities [126]. Directed capital flows to socially responsible market options will sustainably contribute to CSR and SRI trends including diversity and inclusion [23]. Overall, financial markets attuned to social responsibility will lift entire industries onto a more socially conscientious level [120]. As such SRI and corporate social justice are attributed the potential to positively impact the financial markets and create socially attentive market systems that improve the overall standard of living and quality of life for this generation and the following [126].

Socially responsible investors fund ethical causes about which they personally care and refrain from ethical infringements. The integration of personal ethics in their portfolio decision-making and the perception of the investment decisions being in sync with personal protected values lets investors identifying themselves with their choice [63]. The alignment of beliefs and actions evokes identification with investments that grant investors the notion of self-consistency. Self-consistency triggers positive feelings and contributes to the self-enhancement of socially responsible investors [27, 101]. Socially conscientious investors are therefore likely to stay with their choices and continue to align personal economic endeavors with social obligations and societal concerns. SRI leverages into a means of expression of accordance with personal values with societal norms and the wider society, even when market conditions change. Socially responsible corporate conduct attributes long-term perspectives. Socially attentive corporate conduct features sustainability considerations of corporate executives who are mindful of future risks and social impacts of their decision-making.

Long-term viability of corporate conduct is ingrained in CSR practices. CSR grants long-term stability of corporate conduct as for creating a supportive business environment and decreasing the likelihood of stakeholder pressure and litigations risks [55, 68, 110]. When taking rising CSR trends into consideration, SRI offers long-term financial prospects [23, 55, 61]. Socially conscientious investors thereby use SRI as a long-term strategy to contribute to society and SRI becomes a stable and crisis-robust market allocation opportunity [50, 101]. Using equity funds and for the first-time hedge funds, Hong and Jiang [39] show that stocks with high exit rates consistently underperform the market throughout the entire 1980–2008 sample, leading to the conclusion that stability pays [76].

As for longest term allocation preferences, pension funds are another excellent way to allocate financial assets toward the future. In addition, bonds appear to hold potential to save assets for posterity. An interesting novel attempt to couple this financial sustainability strategy with environmental sustainability is proposed by Sachs [98, 99], who advocates for funding today's climate mitigation through an intertemporal fiscal policy mix backed by climate bonds and carbon tax. Bonds are debt investments in which investors loan money to an entity, which borrows the funds for a defined period of time at a variable or fixed interest rate. Bonds are primarily used by companies, municipalities, states, and sovereign governments to raise money and finance a variety of future-oriented long-term projects and activities [58]. This solution appears as real-world relevant means to tap into the worldwide USD 80 trillion bond market in order to fund the incentives to a transition to a sustainable path [125]. Carbon tax will also be introduced. Sharing the costs of climate change aversion between and across generations appears as important strategy to instigate immediate climate change mitigation through incentivizing emission reduction and provide adaptation [73, 79, 80, 82]. Overall this turns climate change burden sharing into a Pareto-improving option over time [73, 86, 87].

Future advancements of the idea of CSR and SRI are enacted in the United States Green New Deal (GND) and the European Union European Green Deal with the European Finance Taxonomy. The United States serves as market solution to implement global environmental governance as "the sum of the many ways individuals and institutions, public and private, manage their common affairs." Based on historical foundations, the underlying economic mechanisms of the

GND and contemporary implementation strategies of the GND are John Maynard Keynes' [48] multiplier effect, which speaks about governmental investments trickling down through all layers of the economy, and the Sustainable Development Goals. The GND spending is targeted at social and green causes fostering concepts such as eco-commerce, environmental enterprise, environmental finance, fiscal environmentalism, green accounting, economy, jobs and trading as well as sustainable energy. The economic policies proposed comprise of fiscal and monetary means, innovation efforts and behavioral changes. Concrete recommendations are given on carbon tax, emissions trading, green bonds, absorbing CO_2 and forestation, insurance policies, intergenerational conscientiousness, engaging portfolio managers, ecotax, environmental pricing reform, environmental tariffs, net metering, Pigovian tax and sustainable tourism. All these efforts are targeted to support global environmental governance.

4.4 Demographics

Demographics can serve as an indicator for future purchasing behavior and hence successful industries. For instance, with the baby boomer generation retiring soon and this population segment being an unproportional holder of wealth, one can estimate retirement wealth spending. Industries like tourism but also retirement leisure activities and health care are prospectively industries that will be prosperous in the coming decade and flourish after the COVID-19 crisis.

In Europe, migration from Middle East may lead to a demand in respective financial products and certain industries that are specialized in Islamic banking, which appears as successful industry in the future years to come.

4.5 Tangibility

Profiting from insights about your heuristics can either occur through saving or making money off your mental limitations and shortcuts. One way to save or cut on your spending is the realization of different spending patterns caused by the tangibility of assets. Credit cards have been found to dangerously raise spending behavior. A direct implication calls for cautious use of credit cards—e.g., lower the amount of credit cards to one or just using credit cards for paying when necessary, such as in foreign sales or electronic transfers. Sharing information on credit card purchases with trusted others may also help by heightened oversight and transparency curbing compulsion.

Market actors tend to have large-scale deficiencies to scale risks and externalities with uncertain outcomes, which are not factored accordingly into market calculus (Hong et al. 2019). For instance, regulatory concerns of markets inexperienced with climate change underreact to such risks and call for disclosing corporate exposures to risks of global warming (Hong et al. 2019).

In addition to this favorable tangibility effect, one may also be aware that the mere presence of wealth may elicit effects in human beings. The behavioral abundance effect shows that unethical behavior emerges in the presence of wealth, potentially through the mechanism envy. Knowing that, in the eye of a large amount of cash, one may feel envy toward others and may compensate by trying to buy or make up for the depleted self-esteem may cause harmful high purchasing decisions, leads to the advice that one should stay away of the visible proximity of monetary wealth and not carry large amounts of money when going shopping in a wealth abundant area.

4.6 Safe Havens

Similar to casino gambling strategies, in order to avoid falling prey to sunk cost fallacy losses, behavioral financiers should divide any gains and always just use a portion of past gains for future bets. The remainder should be secured in a long-term secured savings option. Saving can be enhanced by considering and translating purchase prices in time spent to work for the price (including tax deductions) and an awareness of the scarce amount of time that can never be regained.

4.7 Inflation

"Inflation is the decline of purchasing power of a given currency over time" [26]. This sustained increase in the general level of prices for goods and services in a country occurs when a national money supply growth outpaces economic growth [26]. Inflation, measured as an annual percentage change, indicates that a unit of currency effectively buys less than it did in prior periods [26]. In inflation, prices of goods and services rise over time, purchasing power decreases and people are incentivized to spend money, which boosts the circulation of money and thereby market activities and subsequently economic growth.

Inflation is widely considered to have an inverse relation to unemployment [3, 12]. The Fisher Effect outlines the inverse relation of interest rates and inflation [37]. Central banks around the world engage in the monetary tool of inflation targeting to hold inflation at a certain level, most likely around 2-3% per year, to ensure a continuous standard of living for all market actors [45]. Especially in regards to lead currencies—such as the United States Dollar, the Eurozone Euro, the United Kingdom Pound Sterling, the Swiss Franc, and the Australian Dollar—inflation can have vast impacts on the affordability of goods and services in countries that are pegged to these lead currencies.

Our current economic time in the post-COVID-19 economy features prevalent macroeconomic drivers of inflation in the prevailing central bank money supply boom, demand pull, cost push, and monetary debasement [19]. Around the world, central banks and governments were lowering interest rates after the crisis to produce liquidity across the globe. By-far all-time-high global central bank assets creation reached 31 trillion USD for the first time in 2021 [19]. The U.S. issuance of treasury bonds hit unprecedented high levels, while the U.S. Federal Reserve continued monetizing debt through large quantitative easing programs creating excess money supply adding to inflation [19].

Inflation is caused by an oversupply of money in the economy. Just like any other commodity, the prices of money is determined by supply and demand. If there is too much money supply, the price of money goes down and its purchasing power decreases, leading to the prices of everything else priced getting more expensive.

Built-in inflation may trigger a wage-price spiral, which occurs when workers demand higher wages to keep up with rising living costs. This in turn causes businesses to raise their prices in order to offset their rising wage costs, leading to a self-reinforcing loop of wage and price increases [19]. Demand-pull inflation refers to situations where there are not enough products or services being produced to keep up with demand, causing their prices to increase [19]. Currently, demand is expected to skyrocket after household savings peaked throughout the end of 2020 coupled with consumer spending unleashed throughout 2021 [19]. Trends already project 7.5 times higher the historical average annual spending adjusted for inflation for 2021.

Inflation is also likely caused when corporate costs of production increase, forcing businesses to raise their prices to maintain their profit margins. Production during pandemic constraints has likely increased costs including wages, taxes and natural resources, imports and safety of labor. Supply shortages not only occurred during lockdown. Production disruptions and consumption patterns shifted abruptly in the beginning of the external shock of COVID-19 [34]. Labor shortages also occurred in key qualification frontline workers as well as professions with a vast array of human touchpoints [123]. Pro-active care for maintaining a healthy workforce and the overall long-term well-being of employees, including preventive care in teams, will become an essential corporate feature to attract qualified labor, whose bargaining power increases in the eye of labor shortages in human-facing industries and positions [29, 123]. The COVID long haul prospect—as of the middle of 2022 it is estimated that 7.5% of the U.S. population suffer from COVID-19 long-term health impairments—and prolonged COVID pandemic have risen U.S. job openings to historic highs and the share of U.S. corporations not being able to fill positions has increased to 15%, which accounts for a record highest rate ever measured [29, 116, 123]. At the same time, more workers are quitting with the job quits rate at 2.7% in June 2021 [123]. In addition, the U.S. retail inventory to retail sales ratio has been at an all-time low and labor-management conflicts are expected to rise with the widening of the finance market versus real economy gap likely to becoming even more accentuated in the future [19].

Monetary debasement is another driver of inflation that currently adds inflationary pressure to the U.S. economy featuring a twin deficit of 25% of the nominal GDP [19]. Corporate debt has grown to near all-time highs while bond yields remain near all-time lows [19]. Excess liquidity in the market has created one of the most speculative investment environments of all times with cryptocurrencies and social online media playing a peculiar, novel role in overheating markets. At the same time, commodities such as food and beverage, which already had elevated price levels since 2000 and experienced a significant rise during the 2008/09 World Financial Recession that has not adjusted to previous levels, have exhibited vertical price spikes within the last year [18]. Adding to commodities prices explosions, the aggregate free cash flow for the top 50 gold and silver miners is now growing exponentially [18]. With growing concerns of monetary debasement emerging, investors continue to bid up commodity prices, leading to higher consumer prices [18]. In addition, the S&P 500 earnings yield adjusted for the Consumer Price Index (CPI) is now at its most negative level in 30 years, which has been attributed as an understated measure of true inflation [18].

Understanding the role of the exponential rise in prices for the stability of the underlying economy is key when trying to avoid historical examples of the past, such as the hyperinflation during the interwar period of the first quarter of the twentieth century [38]. International transmissions of inflation are discussed in historic examples to build adaptive expectations on information about current and past experiences with inflation [35, 109]. Disequilibrium inflation occurs when the actual rate of inflation is greater than the expected followed by social unrest and industrial conflict [35, 109]. As a more recent positive example, during the 2008–2009 World Financial Recession, the United States Federal Reserve dropped its interest rate target to close to zero, and then was forced to use unconventional monetary policy tools including quantitative easing as an emergency measure to stimulate the economy and prevent if from tumbling into a deflationary spiral [18, 25]. Treasure bonds issuance was used in order to avoid a free fall inflationary price drop [25]. Due to the treasury bonds issuance and trust in the U.S. economic performance, the U.S. government remained intact as the economy continued an overall productive pace during the period of the Great Recession and only saw very modest increases in inflation [18, 25].

A potential wage-price spiral—such as the one in the 1970s—is likely to be followed in the decades to come, driving investment in raw materials, real-estate and innovative finance options, such as cryptocurrencies [19]. The number of companies mentioning inflation in earnings calls and people searching for the term "inflation" on Google is growing at a pace as never before, which feeds into the general population not feeling confident to hold cash or save money in the bank accounts [19]. This kind of behavior spurs large-scale investment and/or highly risky myopic growth consumption—such as in investments in cryptocurrencies—fueling further bubble build-up potential. All these developments shape economic booms and busts. The underlying market behavior, however, is hardly understood or covered [19, 49, 81). Yet information about the economy is key in shaping collective moods that determine individual market actors' propensity to consume or invest, which leads to specific underlying economic long-term cycles and their positive or negative socio-economic fallout propensities.

While economics offers multiple accounts on the relation of inflation with interest rate dynamics and purchasing behavior shifts in various inflation rate scenarios, less is known about inflationary effects on different social strata. Inflation affects different people in different ways, with some benefiting from its effects at the expense of others who relatively lose out. Inflation therefore can be viewed positively or negatively depending on the individual viewpoint and rate of change.

Obviously, creditors (the lenders of funds) are in a less favorable position with inflation when it comes to the value of money than debtors (the borrowers of funds) economically if considering the overall purchasing power of creditors declines over time, while the overall debt burden for the borrower declines with time. Higher inflation thus harms savers because it erodes the purchasing power of the money they have saved in their accounts and benefits borrowers because the inflation-adjusted value of their outstanding debts shrinks over time [26]. But this effect is a rather economic mechanism, in which creditors lose degrees of freedom over time by decreasing the purchasing power of the funds they expect to receive back. Higher inflation can trigger two scenarios: For one, inflation encourages spending, as consumers will aim to purchase goods quickly before their prices rise further. Savers, on the other hand, could see the real value of their savings erode, limiting their ability to spend or invest in the future [26].

What is missing in economic writing is attention to the psychological account of debtors, who often cannot afford purchases on their own, lock themselves in debt repayment prisons and lower their purchasing power freedom to reallocate funds over time. Never having the financial means to finance large-scale choices flexibly but to constantly having to pay back old debt puts individuals in a backward-looking prison of constant debt repayment of past needs and wants. When people are spending their real economy lives hunting the shadows of their past consumption choice, life goes backwards and rests in the yesterday. Capital being the withheld consumption today grants future flexibility of consumption. Consumption that does not result in debt may grant peace of mind rather than imposing emotionality in the face of long-term constraints [21]. Money at hand thus is psychic energy released at the holder's discretion giving the ultimate freedom of future-oriented choice [21].

In the international arena, inflation is a double-edged sword for import and export industries. Inflation declines a nation's currency's value, which benefits exporters by making their goods more affordable when priced in the currency of foreign nations. Inflation at the same time harms importers by making foreign-made goods more expensive [26]. Thus, inflation is neither good or bad—it more depends on if your revenue or purchasing power are based on imports or exports.

Inflation hits metropolitan populations harder than rural populations, not only in the general price levels but the relative income spent on living expenses. The currently found economic Doughnut effect speaks about richer metropolitan segments having moved out of cities to live in suburbs (hence the Doughnut) during the COVID-19 pandemic, while poorer individuals, who cannot afford relocation costs remain in the city and are more likely to have lost the income sources stemming from richer populations' purchasing power [94].

Inflation effects also depend on whether inflationary changes are anticipated or unanticipated. For instance, when considering that the renegotiation of contracts may take a certain period of time. The question here is not whether inflation occurs but who has flexibility at what pace to renegotiate salaries and other transfer payments for living expense conditions. People living off a fixed-income or social welfare payment beneficiaries—such as those on food stamps, governmental welfare programs, retirees and annuitants, see a sticky—hence hardly or slowly change over time, or non-renegotiable decline in purchasing power that is likely not adjusted for inflation fast enough or at all. People being able to renegotiate their terms or—again—people with savings that can wait out negotiations benefit more from inflation than those who are constraint and pay off old mortgages under tight budget constraints, resulting in decreased negotiation power and likely facing high costs for refinancing or a personal bankruptcy, which may lead to social misery and self-harming choices.

Those with tangible assets—like property or stocked commodities—benefit from moderate inflation as these asset holdings' values are likely to increase with inflation [15, 67, 118]. People holding cash may typically prefer no inflation, as it erodes the value of their cash holdings [26]. Homeowners and landlords rather benefit from inflation, especially if homeownership is built on a fixed interest or no mortgage. Housing prices tend to rise in inflationary markets and homeowners are able to raise the rent each year or lease term [15, 67, 118].

So the discussion of our time should not be about whether the size of the quantity of money established to aid and back economic recovery will likely cause inflation but to whom transfer payments should be redirected in order to offset the negative disparate impact of inflation. It is thus not to label inflation good or bad per se but to rather educating people about their own personal situation and aid in their decisions to navigate in the jungle of finance more strategically in a fast-paced and ever-changing world.

4.8 Interest Rate

Government bailout packages are partially financed via debt and a historically lowest, never-so-long-low key interest rates for savings to spur financial market activity [10, 84]. The U.S., but also other major economies around the world, are expected to face historically low interest rates for savings below growth rates for a long time into the future [10, 84].

Low interest rates allow investors and industry to borrow money for innovative corporate and financial endeavors, which will overall boost the economy and provide a vital liquidity stream for the finance sector [49]. Falling interest rates lubricate the spread of capital across the globe and foster innovative activities by heightening the prospect of a net rate of profit. The economist John Maynard Keynes [49] described the underlying subjective and objective factors of aggregate consumption patterns and factors that influence individual savings propensities.

Subjective factors are the desire to save for future consumption and contingencies, to use passive and speculative investment to expand future income, to amass wealth, and for some, even to enjoy miserliness [49]. Objective factors include windfall gains or losses, taxation, price controls, expectations, and changes in the interest rate [106]. Institutional and organizational factors shape and channel all such factors.

In Keynes' [49] General Theory, the interest rate is subjective and interest rates are linked with each other by maturity [106]. Keynes [49] postulates the interest rate being determined by subjective preferences and the price of a commodity is based on the cost of its production [106]. In Keynes [49], prices and the interest rate are based on expectations how the market will perform, [106]. The government can step in by reducing the interest rate to override the fall in confidence through investment stimulus [49, 106]. A low interest rate regime may crowd out capital accumulation and spur financial investment flows [49, 106].

At the same time, a cut in interest rates also fuels international bubbles in finance and real estate as well as leads to consumers being incentivized to spend rather than save, which puts them at risk of lowered resilience in times of crises [67]. Excessive borrowing may also imply that corporations and governments become excessively indebted, which raises liquidity constraints and market risks. When investing stops and elevated debt levels become endemic, banking liquidity and governmental credit decrease and the economy will enter a system-inherently-induced recession, such as the market downturn in 2008/09. When savings interest rates are low and public debt accumulates, social welfare costs arise in debt rollovers, which crowds out capital accumulation and implies missed welfare opportunities in the future [10].

Low key interest rates will continue to spread the inequality gap of the finance world versus the real economy performance. Low interest rates and access to capital will allow the capital market to flourish based on incentivized liquidity and access to venture capital. Long low interest rate regimes will—at the same time—weaken the potential of the interest rate as a monetary policy tool of central banks. The macroeconomic ineffectiveness of lowering an already low interest rate even further was first described by Keynes [49] as a "liquidity trap" at the beginning of the twentieth century during the Great Depression. The blasé consumer sentiment over a marginal decline of an already low interest rate was also discussed by Ben Bernanke during his term as the Chairman of the Federal Reserve during the 2008/09 Financial Recession.

From the hardly captured socio-psychological standpoint, a low interest rate promotes a rise in consumer debt. With falling interest rates and credit being made easier, consumers and other spending continues to rise. A low interest rate policy thereby brings along long-term external financing of past ideas, which impairs the flexibility of investors to allocate liquidity to future-oriented innovations and may hold back societal progress and thus actually decrease general welfare on the long run [10]. A rising tide of debt also increases the likelihood of real economy constraints and subsequently financial market crashes if people realize that endemic debt levels are no longer sustainable.

The Austrian School of Economics describes excessive issuance of bank credit as a driver of crises in banking systems [102, 103, 104, 105]. Excessive issuance of bank credit is exacerbated if central bank monetary policy sets interest rates too low. The resulting expansion of money supply causes a boom, in which resources are misallocated because of artificially low interest rates and inflation rise. This unsustainable boom is followed by a bust, in which malinvestments are liquidated below their original costs and the money supply contracts and then the financial market system collapses.

Different groups in society may have different outlooks on the same information. While financially versed parts of the population may experience the COVID-19 economic fallout as an opportunity—given modern finance tools and methods such as shorting, derivatives and diversification—the real economy may hit stronger feelings in the eye of uncertainty given the lower levels of degrees of freedom fueled by low interest rates that incentivized debt cycles for consumer credits [53, 56]. What is new to this argument nowadays, is the fact that we are seeing massive socio-psychological burdens in private households these days in the middle class of the United States [14]. Consumer debt tends to create lurking social misery on the long run that has a socio-psychological propensity fallout for suboptimal choices, decisions with negative overall impact over time and self-destructively harmful behavior. A so-coined "deaths of despair" trend is already noticed in the U.S. for mid-career population segments that face death spikes induced by alcoholism, drug use and suicide—all problems that have been exacerbated during the COVID-19 pandemic [14, 16]. Low interest rates on savings accounts in the real economy keep people trapped in the debt financing of past aspirations [4]. Private household debt traps differ from public or governmental debt as the Ricardian equivalent points out in intergovernmental entities' ability to issue bonds and roll over debt to future generations in comparison to private household entities [10]. Financial managers may therefore currently benefit from the low interest regimes in access to capital at the socio-psychological fall-out expense of the real economy and in the eye of an overall market insecurity risk implied in low interest rate regimes on the long run.

One related theory that was brought forward recently was the idea of Sianne Ngai's [66] Gimmicks that give a tableau of illusions that do not add up to reality [14]. Gimmicks include the camouflage of unequal expansion of capitalist gains and economic opportunities by mass media, literature, television, film, video and art to social media online that procreates cultures of aspirations that actually make people unhappy on the long run [66]. Good life fantasies and lifestyle aspirations in a low interest rate climate may create prisons of debt for indebted consumers who spend their whole life paying off bills of previous choices based on past aspirations. Their consumption patterns and financial flows psychologically go backwards in time. This cruelty in the optimism portrayed in media created by capitalist machineries may be associated to the so-called death of despair—rising numbers of self-induced deaths in the young middle class through such means as alcohol, drug intake and suicide noticeable in today's capitalist society [7, 14, 16, 66].

References

1. Agrawal, G., Ahlawat, H. & Dewhurst, M. (2020). Winning against COVID-19: The implications for biopharma, McKinsey & Co., April 15, 2020. Retrieved at https://www.mckinsey.com/industries/pharmaceuticals-and-medical-products/our-insights/winning-against-covid-19-the-implications-for-biopharma#the-imp.
2. Alperson, M., Tepper-Marlin, A. T., Schorsch, J., & Wil, R. (1991). *The better world investment guide: One hundred companies whose policies you should know about before you invest your money. From the Council on Economic Priorities.* Prentice Hall.
3. Armstrong, Ph., Glyn, A., & Harrison, J. (1991). *Capitalism since 1945.* Blackwell.
4. Arora, R. (2020). Which companies did well during the Coronavirus pandemic? *Forbes*, June 30, 2020. Retrieved at https://www.forbes.com/sites/rohitarora/2020/06/30/which-companies-did-well-during-the-coronavirus-pandemic/?sh=418c3bda7409
5. Ashraf, N., Karlan, D. S., & Yin, W. (2006). Tying Odysseus to the mast: Evidence from a commitment savings product in the Philippines. *Quarterly Journal of Economics, 121*(2), 635–672. https://doi.org/10.1162/qjec.2006.121.2.635
6. Beltratti, A. (2003). *Socially responsible investment in general equilibrium: Economic theory and applications.* Retrieved from www.ssrn.com
7. Berlant, L. G. (2011). *Cruel optimism.* Duke University Press.
8. Beshears, J. L., Choi, J. J., Laibson, D. I., Madrian, B. C., & Sakong, J. (2011). *Self-control and liquidity: How to design a commitment contract.* RAND Working Paper Series, WR- 895-SSA. Retrieved at SSRN: http://ssrn.com/abstract=1970039 or http://dx.doi.org/10.2139/ssrn.1970039
9. Biller, A. (2007). Socially responsible investing now part of the landscape. *Benefits and Compensation Digest, 44,* 12.
10. Blanchard, O. (2017). Public debt and low interest rates. *American Economic Review*, 109(4), 1197–1229.
11. Bollen, N. P. B., & Cohen, M. A. (2004). *Mutual fund attributes and investor behavior.* Unpublished working paper, Vanderbilt University.
12. Brenner, R. (2003). *The boom and the bubble: The US in the World Economy.* Verso.
13. Broadhurst, D., Watson, J., & Marshall, J. (2003). *Ethical and socially responsible investment. A reference guide for researchers.* Saur.
14. Case, A. & Deaton, A. (2020). *Death of despair and the future of capitalism.* Princeton University Press.
15. Cassim, Z., Handjiski, B., Schubert, J., & Zouaoui, Y. (2020). *The 10 trillion rescue: How governments can deliver impact Governments have announced the provision of trillions of dollars in crisis relief, but translating that into sustained recovery will not be easy.* McKinsey Public Sector Practice Report, June 2020. Retrieved at https://www.mckinsey.com/~/media/McKinsey/Industries/Public%20Sector/Our%20Insights/The%2010%20trillion%20dollar%20rescue%20How%20governments%20can%20deliver%20impact/The-10-trillion-dollar-rescue-How-governments-can-deliver-impact-vF.pdf
16. CDC Center for Disease Control and Prevention. (2020). *Symptoms of Coronavirus Disease, Coronavirus Disease 2019 (COVID-19).* https://www.cdc.gov/coronavirus/2019-ncov/symptoms-testing/symptoms.html
17. Chabris, Ch., Laibson, D., & Schuldt, J. (2008). *Intertemporal choice.* Palgrave Dictionary of Economics.
18. Cox, J. (2021). Consumer prices jump 5% in May, fastest pace since the summer of 2008, *CNBC News*, June 10, 2021. Retrieved at https://www.cnbc.com/2021/06/10/cpi-may-2021.html
19. Crescat Capital. (2021). *May research letter: The 3 pillars of inflation*, May 19, 2021. Retrieved at https://www.crescat.net/may-research-letter/

20. Crane, A., & Livesey, S. (2002). Are you talking to me? Stakeholder communication and the risks and rewards of dialogue. In J. Andriof, S. Waddock, S. Rahman, & B. Husted (Eds.), *Unfolding stakeholder thinking* (pp. 39–52). Greenleaf.
21. Csikszentmihalyi, M. (2003). *Good business: Leadership, flow, and the making of meaning*. Viking.
22. Dhrymes, P. J. (1998). *Socially responsible investment: Is it profitable? The investment research guide to socially responsible investing*. The Colloquium on Socially Responsible Investing. Retrieved from http://www.columbia.edu/~pjd1/
23. Dupré, D., Girerd-Potin, I., & Kassoua, R. (2008). *Adding an ethical dimension to portfolio management*. Working paper. Retrieved from www.ssrn.com
24. DW, Deutsche Bank überrascht mit hohem Gewinn, DW.com, October 28, 2020. Retrieved at https://www.dw.com/de/deutsche-bank-%C3%BCberrascht-mit-hohem-gewinn/a55417971.
25. Federal Reserve Bank of St. Louis. (2015). *The origins of unconventional monetary policy in the U.S.*. Retrieved at https://www.stlouisfed.org/annual-report/2015/the-origins-of-unconventional-monetary-policy-in-the-us
26. Fernando, F. (2021). What Is inflation? *Investopedia*, June 17, 2021. Retrieved at https://www.investopedia.com/terms/i/inflation.asp
27. Frey, D., & Irle, M. (2002). *Theorien der Sozialpsychologie*. Hans Huber.
28. Geczy, Ch. C., Stambaugh, R. F., & Levin, D. (2003). *Investing in socially responsible mutual funds*. The Wharton School.
29. Ghilarducci, T. (2021). Are firms causing people to quit? Quits rate increasing to 2.7%, *Forbes*, August 14, 2021. Retrieved at https://www.forbes.com/sites/teresaghilarducci/2021/08/14/are-firms-causing-people-to-quit-quits-rate-increasing-to-27/?sh=3a18e42d35d0
30. Gill, A. (2001). *Saints and sinners: Who's got religion?*. Credit Lyonnais Securities Asia Report.
31. Gine, X., Karlan, D., & Zinman, J. (2009). *Put your money where your butt is: A commitment contract for smoking cessation*. Policy Research Working Paper Series 4985, The World Bank.
32. Gourville, J. T., & Soman, D. (2005). Overchoice and assortment type: When and why variety backfires. *Marketing Science, 24*(3), 295–382. https://doi.org/10.1287/mksc.1040.0109
33. Guenster, N., Derwall, J., Bauer, R., & Koedijk, K. (2005). *The economic value of corporate eco-efficiency*. Unpublished working paper, Erasmus University.
34. Helper, S., & Soltas, E. (2021). *Why the pandemic has disrupted supply chains*. The White House of the United States, June 17, 2021. Retrieved at https://www.whitehouse.gov/cea/written-materials/2021/06/17/why-the-pandemic-has-disrupted-supply-chains/
35. Hall, P. A., & Soskice, D. (2001). An introduction to varieties of capitalism. In P. A. Hall & D. Soskice (Eds.), *Varieties of Capitalism: The Institutional Foundations of Comparative Advantage* (pp. 8–9). Oxford University Press.
36. Harvey, C. (2008). *Campbell R. Harvey's hypertextual finance glossary*. Retrieved http://www.duke.edu/charvey/Classes/wpg/bfglosa.htm
37. Hayes, A. (2020). Fisher effect definition. *Investopedia*, August 23, 2020. Retrieved at https://www.investopedia.com/terms/f/fishereffect.asp
38. Hayes, A. (2021). Why didn't quantitative easing lead to hyperinflation? *Investopedia*, June 27, 2001. Retrieved at https://www.investopedia.com/articles/investing/022615/why-didnt-quantitative-easing-lead-hyperinflation.asp
39. Hong, H. G., & Jiang, W. (2011). *When some investors head for the exit*. American Finance Association (AFA), 2013 San Diego Meetings Paper.
40. Hong, H. G., & Kacperczyk, M. (2007). The price of sin: The effects of social norms on markets. *Journal of Financial Economics, 93*(1), 15–36. https://doi.org/10.1016/j.jfineco.2008.09.001
41. Hong, H. G., & Liskovich, I. (2016). *Crime, punishment and the value of corporate social responsibility*. Retrieved from https://www8.gsb.columbia.edu/programs/sites/programs/
42. Hong, H. G., & Yogo, M. (2012). What does futures market interest tell us about the macroeconomy and asset prices? *Journal of Financial Economics, 105*(3), 473–490. https://doi.org/10.1016/j.jfineco.2012.04.005

43. Houser, D., Schunk, D., Winter, J., & Xiao, E. (2010). *Temptation and commitment in the laboratory*. IEW Working Paper iewwp 448, Institute for Empirical Research in Economics, University of Zurich.
44. Irwin, J. R., Slovic, P., Lichtenstein, S., & McClelland, G. (1993). Preference reversals and the measurement of environmental values. *Journal of Risk and Uncertainty, 9*, 5–38. https://doi.org/10.1007/BF01065347
45. Jahan, S. (2021). *Inflation targeting: Holding the line*. International Monetary Fund, December 8, 2021. Retrieved at https://www.imf.org/external/pubs/ft/fandd/basics/target.htm
46. Kahneman, D., & Ritov, I. (1994). Determinants of stated willingness to pay for public goods: A study in the headline method. *Journal of Risk and Uncertainty, 9*, 5–38. https://doi.org/10.1007/BF01073401
47. Kaur, S., Kremer, M., & Mullainathan, S. (2010). Self-control and the development of work arrangements. *American Economic Review, 100*(2), 624–628. https://doi.org/10.1257/aer.100.2.624
48. Keynes, J. M. (1936). *The general theory of employment, interest and money*. Macmillan.
49. Keynes, J. M. (1936/2003). *The general theory of employment, interest and money*. Harvard University Press.
50. Knoll, M. S. (2008). *Socially responsible investment and modern financial markets*. Unpublished working paper, University of Pennsylvania Law School.
51. Laibson, D. I. (1997). Golden eggs and hyperbolic discounting. *Quarterly Journal of Economics, 11*, 443–477. https://doi.org/10.1162/003355397555253
52. Laibson, D. I., Repetto, A., & Tobacman, J. (2003). A debt puzzle. In P. Aghion, R. Frydman, J. Stiglitz & M. Woodford (Eds.), *Knowledge, information, and expectations in modern economics: In honor of Edmund S. Phelps* (pp. 228–266). Princeton University Press.
53. Lee, B., & Martin, R. (2016). *Derivatives and the wealth of societies*. University of Chicago Press.
54. Lerner, S. (2020). Big pharma prepares to profit from the Coronavirus: Pharmaceutical companies view the Coronavirus pandemic as a once-in-a-life time business opportunity, The Intercept, March 13, 2020. Retrieved at https://theintercept.com/2020/03/13/bigpharma-drug-pricing-coronavirus-profits/.
55. Little, K. (2008). *Socially responsible investing: Put your money where your values are*. Penguin.
56. LiPuma, E., & Lee, B. (2004). *Financial derivatives and the globalization of risk*. Duke University Press.
57. Markowitz, H. M. (1959). *Portfolio selection: Efficient diversification of investments*. Wiley.
58. Marron, D. B., & Morris, A. C. (2016). *How to use carbon tax revenues*. Tax Policy Center Urban Institute & Brookings Institution. https://doi.org/10.2139/ssrn.2737990
59. McClure, S., Ericson, K., Laibson, D., Loewenstein, G., & Cohen, J. (2007). Time discounting for primary rewards. *Journal of Neuroscience, 27*(21), 5796–5804. https://doi.org/10.2139/ssrn.2737990
60. McWilliams, A., & Siegel, D. (2000). Corporate social responsibility and financial performance: Correlation or mis-specification? *Strategic Management Journal, 21*, 603–609. https://doi.org/10.1002/(SICI)1097-0266(200005)21:5<603::AID-SMJ101>3.0.CO;2-3
61. McWilliams, A., Siegel, D., & Teoh, S. W. (1999). Issues in the use of the event study methodology: A critical analysis of corporate social responsibility studies. *Organizational Research Methods, 2*, 340–365. https://doi.org/10.1177/109442819924002
62. Meier, S., & Sprenger, C. (2010). *Stability of time preferences*. IZA Discussion Paper 4756, Institute for the Study of Labor.
63. Mohr, L. A., & Webb, D. J. (2005). The effects of corporate social responsibility and price on consumer responses. *Journal of Consumer Affairs, 39*(1), 121–147. https://doi.org/10.1111/j.1745-6606.2005.00006.x
64. Mohr, L. A., Webb, D. J., & Harris, K. E. (2001). Do consumers expect companies to be socially responsible? The impact of corporate social responsibility on buying behavior. *Journal of Consumer Affairs, 35*(1), 45–72. https://doi.org/10.1111/j.1745-6606.2001.tb00102.x

65. Myers, St. (1984). The capital structure puzzle. *Journal of Finance, 39*, 575–592. https://doi.org/10.2307/2327916
66. Ngai, S. (2020). *Theory of the Gimmick: Aesthetic judgment and capitalist form*. Harvard University Press.
67. Passy, J. (2021). An inflation storm is coming for the U.S. housing market, *MarketWatch*, June 11, 2021. Retrieved at https://www.msn.com/en-us/money/realestate/an-inflation-storm-is-coming-for-the-u-s-housing-market/ar-AAKWtdx?ocid=msedgntp
68. Posnikoff, J. F. (1997). Disinvestment from South Africa: They did well by doing good. *Contemporary Economic Policy, 15*(1), 76–86. https://doi.org/10.1111/j.1465-7287.1997.tb00456.x
69. Puaschunder, J. M. (2010). *On corporate and financial social responsibility*. Unpublished Doctoral Thesis. University of Vienna, Faculty of Psychology.
70. Puaschunder, J. M. (2013). Ethical investing and socially responsible investing. In K. H. Baker & V. Ricciardi, (Eds.), *Investor behavior* (pp. 515–532). Wiley Finance Series.
71. Puaschunder, J. M. (2015a). The call for global responsible intergenerational leadership in the corporate world: The quest of an integration of intergenerational equity in contemporary Corporate Social Responsibility (CSR) models. In *Proceedings of the 2015 6th International Conference of the Association of Global Management Studies at Alfred Lerner Hall of Columbia University*. The Association of Global Management Studies.
72. Puaschunder, J. M. (2015b). When investors care about politics: A meta-synthesis of political divestiture studies on the capital flight from South Africa during Apartheid. *Business, Peace and Sustainable Development, 5*(24), 29–52. https://doi.org/10.9774/GLEAF.8757.2015.ju.00004
73. Puaschunder, J. M. (2016a). Intergenerational climate change burden sharing: An economics of climate stability research agenda proposal. *Global Journal of Management and Business Research: Economics and Commerce, 16*(3), 31–38.
74. Puaschunder, J. M. (2016b). On eternal equity in the fin-de-millénaire: Rethinking capitalism for intergenerational justice. *Journal of Leadership, Accountability and Ethics, 13*(2), 11–24.
75. Puaschunder, J. M. (2016c). On the emergence, current state and future perspectives of socially responsible investment (SRI). *Consilience: The Columbia University Journal of Sustainable Development, 16*(1), 38–63.
76. Puaschunder, J. M. (2016d). Socially responsible investment as emergent risk prevention and means to imbue trust in the post-2008/2009 world financial crisis economy. In O. Lehner (Ed.), *Routledge handbook of social and sustainable finance* (pp. 222–238). Taylor and Francis.
77. Puaschunder, J. M. (2016e). The call for global responsible inter-generational leadership: The quest of an integration of inter-generational equity in corporate social responsibility (CSR) models. In D. Jamali (Ed.), *Comparative perspectives on global corporate social responsibility* (pp. 276–289). IGI Global Advances in Business Strategy and Competitive Advantage Book Series.
78. Puaschunder, J. M. (2016f). The role of political divestiture for sustainable development. *Journal of Management and Sustainability, 6*(1), 76–91. https://doi.org/10.5539/jms.v6n1p76
79. Puaschunder, J. M. (2017a). Financing climate justice through climate change bonds. *Oxford Journal of Finance and Risk Perspectives, 6*(3), 1–10.
80. Puaschunder, J. M. (2017b). Krisenrobuste alternativen. *Global Investor, 3*(5), 69.
81. Puaschunder, J. M. (2017c). Socio-psychological motives of socially responsible investors. *Global Corporate Governance: Advances in Financial Economics, 19*, 209–247.
82. Puaschunder, J. M. (2017d). The climatorial imperative. *Agriculture Research and Technology, 7*(4), 1–2.
83. Puaschunder, J. M. (2018a). *Corporate social responsibility and opportunities for sustainable financial success*. IGI Publishing.
84. Puaschunder, J. M. (2018b). Socio-psychological motives of socially responsible investments (SRI). In S. Boubaker, D. Cumming, & D. K. Nguyen (Eds.), *Research handbook of investing in the triple bottom line: Finance, society and the environment* (pp. 447–472). Edward Elgar.

85. Puaschunder, J. M. (2018c). The history of ethical, environmental, social and governance-oriented investments as a key to sustainable prosperity in the finance world. In S. Boubaker & D. K. Nguyen (Eds.), *Corporate social responsibility, ethics and sustainable prosperity* (pp. 359–388). World Scientific.
86. Puaschunder, J. M. (2019a). An inquiry into the nature and causes of Climate Wealth of Nations: What temperature finance gravitates towards? Sketching a climate-finance nexus and outlook on climate change-induced finance prospects. *Archives of Business Research, 7*(3), 183–217.
87. Puaschunder, J. M. (2019b). Intergenerational governance and leadership around the world. In J. M. Puaschunder (Ed.), *Intergenerational governance and leadership in the corporate world* (pp. 153–177). Idea Group.
88. Puaschunder, J. M. (2019c). Lessons from divesting South Africa during Apartheid for a green economy. In A. Tavidze (Ed.), *Progress in economics research* (pp. 119–147). Nova.
89. Puaschunder, J. M. (2019d). Long-term investments. In W. L. Filho et al. (Ed.), *Partnerships for the goals: Encyclopedia of the UN sustainable development goals,* pp. 108–118. Springer Nature.
90. Puaschunder, J. M. (2019e). When investors care about causes: Learning from political divestiture during Apartheid and the Sudan crisis for the transition to renewable energy via green bonds. In F. Lukhele (Ed.), *Apartheid in South Africa: Reflections and lessons* (pp. 1–28). Nova.
91. Puaschunder, J. M., Gelter, M., & Sharma, S. (2020). COVID-19 shock: Socio-technical, legal, corporate, economic and governance changes, Research Association for Interdisciplinary Studies Conference on Social Sciences & Humanities Proceedings, August 22, 2020, pp. 82–93.
92. Puaschunder, J. M. (2022). *Ethics of inclusion: The cases of health, economics, education, digitalization and the environment in the post-COVID-19 era.* Ethics International.
93. Puaschunder, J. M., & Schwarz, G. (2012). *The future is now: How joint decision-making curbs hyperbolic discounting but blurs social responsibility in the intergenerational equity public policy domain.* Harvard University Situationist Law and Mind Sciences Working Paper.
94. Ramani, A., & Bloom, N. (2021). *The donut effect of Covid-19 on cities.* National Bureau of Economic Research Working Paper No. 28876. https://www.nber.org/papers/w28876
95. Read, D., Loewenstein, G., & Kalyanaraman, S. (1999). Mixing virtue and vice: The combined effects of hyperbolic discounting and diversification. *Journal of Behavioral Decision Making, 12,* 257–273. https://doi.org/10.1002/(SICI)1099-0771(199912)12:4<257::AID-BDM327>3.0.CO;2-6
96. Read, D., & van Leeuwen, B. (1998). Predicting hunger: The effects of appetite and delay on choice. *Organizational Behavior and Human Decision Processes, 76*(2), 189–205. https://doi.org/10.1006/obhd.1998.2803
97. Renneboog, L. D. R., Horst, J. R. T., & Zhang, C. (2007). *Socially responsible investments: Methodology, risk and performance.* Tilburg University Center for Economic.
98. Sachs, J. D. (2007). *Common wealth: Economics for a crowded planet.* Penguin.
99. Sachs, J. D. (2014). Climate change and intergenerational well-being. In L. Bernard & W. Semmler (Eds.), *The Oxford handbook of the macroeconomics of global warming* (pp. 248–259). Oxford University Press.
100. Schroeder, M. (2003). *Socially responsible investments in Germany, Switzerland, and the United States.* Centre for European Economic Research Discussion Paper 03-10, March 2003.
101. Schueth, S. (2003). Socially responsible investing in the United States. *Journal of Business Ethics, 43,* 189–194. https://doi.org/10.1023/A:1022981828869
102. Schumpeter, J. A. (1934). *The theory of economic development.* Harvard University Press.
103. Schumpeter, J. A. (1943/1976). *Capitalism, socialism and democracy.* Allen & Unwin.
104. Schumpeter, J. A. (1949). Economic theory and entrepreneurial history. In R. R. Wohl (Ed.), *Change and the Entrepreneur: Postulates and the Patterns for Entrepreneurial History* (pp. 131–142). Harvard University Press.

105. Schumpeter, J. A. (1989). *Essays on entrepreneurs, innovations, business cycles, and the evolution of capitalism*. Routledge.
106. Shaikh, A. (2016). *Capitalism: Competition, conflict, and crises*. Oxford University Press.
107. Shui, H., & Ausubel, L. M. (2004). *Time inconsistency in the credit card market*. Social Science Research Network working paper. Retrieved Oct 25, 2015, from http://ssrn.com/abstract=586622
108. Siegel, D. S., & Vitaliano, D. F. (2006). *An empirical analysis of the strategic use of corporate social responsibility*. Working paper, Rensselaer Polytechnic Institute.
109. Soskice, D. (1978). Strike waves and wage explosions, 1968–1970: An economic interpretation. In C. Crouch & A. Pizzorno (Eds.), *The Resurgence of Class Conflict in Western Europe since 1968* (pp. 221–245). Macmillan.
110. Sparkes, R. (2002). *Socially responsible investment: A global revolution*. Wiley.
111. Sparkes, R., & Cowton, Ch J. (2004). The maturing of socially responsible investment: A review of the developing link with corporate social responsibility. *Journal of Business Ethics, 52,* 45–57. https://doi.org/10.1023/B:BUSI.0000033106.43260.99
112. Starr, M. (2008). Socially responsible investment and pro-social change. *Journal of Economic Issues, 42*(1), 51–73. https://doi.org/10.1080/00213624.2008.11507114
113. Steurer, R. (2010). The role of governments in corporate social responsibility: Characterising public policies on CSR in Europe. *Policy Science, 43,* 49–72. https://doi.org/10.1007/s11077-009-9084-4
114. Stiglitz, J. E. (2003). *Globalization and its discontents*. Norton.
115. Stone, B. K., Guerard, J. B., Gületkin, M. N., & Adams, G. (2001). *Socially responsible investment screening: Strong evidence of no significant cost for actively managed portfolios*. Working paper. Retrieved from http://www.socialinvest.org/pdf/research/Moskowitz/2001%20Honorable%20Mention%20-%20Moskowitz.pdf
116. Tackett, M. (2022). 7.5% of American population struggles with long COVID, *Aspen Daily News*, June 24, 2022.
117. Thaler, R. H., & Shefrin, H. M. (1981). An economic theory of self-control. *Journal of Political Economy, 89,* 392–406. https://doi.org/10.1086/260971
118. The White House of The United States of America. (2021). Retrieved at https://www.whitehouse.gov/briefing-room/legislation/2021/01/20/president-biden-announces-american-rescue-plan/
119. Tippet, J. (2001). Performance of Australia's ethical funds. *Australian Economic Review, 34*(2), 170–178. https://doi.org/10.1111/1467-8462.00186
120. Trevino, L. K., & Nelson, K. A. (2004). *Managing business ethics: Straight talk about how to do right*. Wiley.
121. Trope, Y., & Fishbach, A. (2004). Going beyond the motivation given: Self-control and situational control over behavior. In R. Hassin, J. S. Uleman, & J. W. Bargh (Eds.), *The new unconscious* (pp. 537–565). Oxford University Press.
122. Tversky, A., & Shafir, E. (1992). Choice under conflict: The dynamics of deferred decision. *Psychological Science, 3*(6), 358–361. https://doi.org/10.1111/j.1467-9280.1992.tb00047.x
123. Whelan, R. (2021). High pay for Covid-19 nurses leads to shortages at some hospitals: Some hospitals pay signing bonuses of up to 40,000; 'We're desperate for nurses,' *The Wall Street Journal*, August 30, 2021. Retrieved at https://www.wsj.com/articles/germany-boosts-already-hefty-coronavirus-stimulus-11598440184
124. Williams, G. (2005). *Are socially responsible investors different from conventional investors? A comparison across six countries*. Unpublished working paper, University of Bath.
125. World Bank. (2015). *World Bank 2015 Report*. World Bank.
126. Zheng, L. (2020). We're entering the age of Corporate Social Justice. *Harvard Business Review*, June 15. Retrieved at https://hbr.org/2020/06/were-entering-the-age-of-corporate-social-justice

5 Market Communication

Behavioral finance is one of the most novel developments in behavioral economics. Since the end of the 1970s, a wide range of psychological, economic, and sociological laboratory and field experiments proved human beings deviating from rational choices and standard neoclassical profit maximization axioms that fail to explain how human actually behave. Human beings were rather found to use heuristics in the day-to-day decision-making. These mental short cuts enable to cope with information overload in a complex world. Behavioral economists proposed to nudge and wink citizens to make better choices for them with many different applications in very many different domains. This chapter reviews and proposes how to use mental heuristics, biases, and nudges in the finance domain to profit from markets with particular attention to the role of information.

5.1 Too Much Information

The fear of too much movement: First, the speed of market communication may manipulate purchases of stocks. Countercyclical communication times may help avoid people falling for options without thinking twice. A lower frequency of communication may help calm jumpy investors, e.g., a tactic used during the EU Brexit referendum. For instance, some major investment companies reacted to the Brexit referendum outcome by doing fortnightly reporting in order to slow people to exit their portfolio with the hope to avoid further market turbulence and to rebalance the shock.

The fear of too much information also entails what kind of information is shared and the credibility of sources. It is important to focus on information that is timely and reliable and abandon incredible resources for making market decisions. Whose information should be trusted in financial markets? Theories of reputation and herd behavior suggest that herding among young career novices is more common than with their more matured counterparts [19, 37, 41]. Security analysts are more likely

J. Puaschunder, *Advances in Behavioral Economics and Finance Leadership*, Contributions to Economics, https://doi.org/10.1007/978-3-031-15710-3_5

to be terminated for inaccurate forecasts than experienced counterparts, leading novices to be less likely to deviate from the expected. Young early career analysts' career concerns causing inexperienced analysts to fall for herding behavior lead to the conclusion to rather trust older, wiser, and more mature analysts [23, 37, 41].

Strategic trading depends on past market information [24]. Past price development impacts perceived uncertainty of prices as well as the equilibrium feedback to prices [24].

Cautious investors should be aware of overconfidence in markets. Analysts who are optimistic relative to the market consensus are more likely to experience favorable career moves [17]. This overconfidence bias of brokerage houses rewarding optimistic analysts exaggerates during market downturns [17].

With growing digitalization and quickening of transfer speed, information exchange in the individual involvement to break trends online on a global scale may impose unknown systemic risks in causing social volatility in international economics. Research may explore how human beings' communication and interaction results in socially constructed volatility that echoes in economic correlates. This paper theoretically covers the history of heterodox economic cycles in order to then propose to explore the role of communication and temporal foci in pandemic communication to create social volatility underlying economic downturns.

5.2 Too Little Information

The fear of too little movement: As outlined by Hong and Stein [27], there is underreaction to information on markets, leading to momentum trading opportunities in markets. In general, stock markets react with delay to information contained in industry returns about their fundamentals and that information diffuses only gradually across markets [30]. Underreaction mainly occurs in the aftermath of crises or when there is a lot of uncertainty in a market. While in the Brexit referendum markets at first, there was the fear of extreme decisions and the speed of the market was slowed, in its aftermath certain markets stopped trading at all. For instance, the housing market in London slowed and almost stopped as for the uncertainty which outcome Brexit may hold in fear of an uncertain future. Or during the Coronavirus crisis, markets at first were spooked and slowed down during the initial lockdown period but later financial markets rebounded back even stronger than ever before. In times of economic crises, the demand for speeding up the economy may arise. Countercyclical policies may aid to speed up the velocity of money within the overall economy.

Information gathering breeds discipline within financial markets and lowers overconfidence biases [10]. Security analyst coverage disciplines credit rating agencies and leads to a drop-in optimism bias in credit ratings, especially for firms with little bond analyst coverage and for firms that are close to default. This coverage-induced shock leads to less informative ratings about future defaults and downgrades and more subsequent bond security mispricings. Even though analysts do not directly compete with credit rating agencies, analyst reports about a firm's equity discipline what credit rating agencies can say about the firm's debt [10].

Social online media, such as Twitter, have gained increasing importance for market communication about highly volatile market instruments—such as cryptocurrencies—in recent months. The trend speaks for an acceleration of financial market communication and democratized access to market information. However, there is also market manipulation and much cognitive noise and mis- and disinformation but also irrational exuberance heating up the overall economy associated with economic market communication on social media.

5.3 Social Phenomenon and Leaders in the Field

While news play some role in determining stock market changes, large market moves often occur on days without any identifiable major news releases. Stock price movements are not fully explicable by news about future cash flows and discount rates [8]. The standard approach holds that fluctuations in asset prices are attributable to changes in fundamental values.

The event study literature has demonstrated that share prices react to announcements about corporate control, regulatory policy, and macroeconomic conditions that plausibly affect fundamentals. Several recent studies of asset pricing have challenged the view that stock price movements are wholly attributable to the arrival of new information. Roll's [36] analysis of price fluctuations in the market for orange juice futures suggests that news about weather conditions, the primary determinant of the price of the underlying commodity, can explain only a small share of the variation in returns. Shiller's [38] claim that stock returns are too variable to be explained by shocks to future cash flows, or by plausible variation in future discount rates, is also an argument for other sources of movement in asset prices.

Frankel and Meese [11] report similar difficulties in explaining exchange rate movements. French and Roll [12] demonstrate that the variation in stock prices is larger when the stock market is open than when it is closed, even during periods of similar information release about market fundamentals. Roll [36] estimated the fraction of return variation that can be attributed to various news, which explains about one-third of the variance in stock returns. Neiderhoffer [35] analyzes stock market reactions to identifiable world news. While news regarding wars, the Presidency, or significant changes in financial policies affects stock prices, the results render it implausible that qualitative news can account for all of the return components that cannot be traced to macroeconomic innovations [8].

Stock market price expectations develop from news, word-of-mouth, and social information sharing. News released to too many leads to an expected diffusion rate. But news initially released to only few leads to an expected diffusion rate that initially increases in time and only then decreases. The serial correlation of stock returns and trading volume are proportional to the diffusion rate [13].

Chen et al. [6] show that entry of investors that have not previously owned the stock is associated with more over-pricing. The exit rate better captures the disagreement distribution of investors in similar fund styles actively evaluating a stock.

Diversity of opinions among investors plays a crucial role in models of financial market speculation and bubbles. By using data from China, it was found that investors living in linguistically diverse areas express more diverse opinions on stock message boards and trade stocks more actively. Language barriers slow news diffusion [4].

Stock market participation is a social phenomenon and therefore highly dependent on the social reference group [20]. Economic bubbles develop out of overconfidence, which is rewarded in markets [25, 28]. Prices drop on the lock-up expiration date [21]. Trading volume appears to be an indicator of sentiment [30]. Bubbles seem to build up slowly based on word-of-mouth recommendations. Market crashes or market downturns are significantly related to days of high trading as investors tend to depend on and learn from signals of their commemorates [30, 28].

Investor relations are established and maintained for the sake of liquidity securitization [14]. Latent social networks of investors play a role in their investment choices, such stock holdings to investors' linkages but also university alumni connections in that city play a role [34]. Social influence also becomes visible in political party affiliations' and presidential election outcomes' impact on financial markets. Party affiliations help predict investment behavior insofar as Democratic investors are more likely to hold or engage in SRI funds than Republican citizens and institutional investors [16]. Republican president elections lead to a market up, while President assassinations and entrance of the market country into war in general lead to a market down [8]. Political events around the world shape the U.S. market [8].

Macroeconomic news explain about one-fifth of movements of stock prices. Most of the macroeconomic news variables affect returns with their predicted signs and statistically significant coefficients [8]. Political development determine future policy expectations and international events influence risk premia, which set pricing. Stock markets react to major non-economic events such as elections, international conflicts, environmental catastrophes and pandemics [8, 35].

A sample of events derived from the Chronology of Important World Events from the World Almanac identified if news reports coincident with stock prices [8]. Some of the events are clearly associated with substantial movements in the aggregate market. On the Monday after President Eisenhower's heart attack in

September 1955, for example, the market declined by 6.62%. The Monday after the Japanese attack on Pearl Harbor witnessed a market decline of 4.37%. The orderly presidential transition after President Kennedy was assassinated coincides with a 3.98% market uptick, while the actual news of the assassination reduced share values by nearly 3%. For the set of events analyzed, the average absolute market move is 1.46%, in contrast to 0.56% over the entire 1941–1987 period [8].

Certain market industries are predictors of the overall stock market performance. In the U.S. and eight other international markets, a significant number of industry returns, including retail, services, commercial real estate, metal, and petroleum, can help forecast the stock market by up to two months [30].

Arbitrageurs tend to amplify economic shocks insofar as speculators holding short positions switch options making highly shorted stocks excessively sensitive to shocks compared to stocks with little short interest [18]. The price of highly shorted stocks overshoots after good earnings news due to short covering compared to other stocks [18].

Observations were reported that many of the largest market movements in recent years have occurred on days when there were no major news events. Further understanding of asset price movements requires model price movements as function of evolving consensus opinions about the implications of given pieces of information. Propagation mechanisms that explain why shocks with small effects on discount rates or cash flows may have large effects on prices. How social representations and collective market movements change in public discourse—and are primarily influenced by online social media—is one of the most exciting research questions in the digital age. This might explain why shocks with smaller impacts on discount rates or cash flows have large impacts on prices. Benevolence of the subconscious wisdom of markets and the organic whole of the economy embodied in the existing social, economic, and legal institutions should be explored. The collective soul of markets embedded in organic aggregate economies anchored in existing social, economic and legal institutions deserves further study.

5.4 Time of Information

The time of information release plays a role as behavioral finance has found. Trading and returns vary in periods of market closure leading to time variations in equilibrium returns [31]. There is more trading activity around close and open and more volatility open-to-open returns than close-to-close returns. This effect reflects information piling until the opening of markets that investors want to react to.

Firm announcements on a Friday lead to a less positive outcome on the market than any other days of the week [9]. The interpretation is that investors are distracted by the weekend and partially forget about the implications of the news [29]. In addition, trading volume is lowered during summer months, assumed due to investors being on vacation [33]. There is a so-called January effect, showing that stock prices tend to rise in January, particularly the prices of small firms and firms whose stock price has declined substantially over the past few years. Also, risky stocks earn most of their risk premiums in January [39]. Overall weekend, holiday, and turn of the month and intraday effects show trading patterns follow calendars [40].

5.5 Firm-Biased Information

Price efficiency plays an important role in financial markets. Firms influence it, particularly when they issue public equity. They can hire a reputable underwriter with a star analyst to generate public signals about profits to reduce uncertainty and increase valuations [2]. These means have a signal function that shapes a favorable image transfer, which tends to raise the price of an option.

5.6 Medium Bias

Stock markets appear to react to certain media stronger than others. Events, which the New York Times carried as lead story and which the New York Times Business Section reported, are having a significant effect on stock market participants. Political, military, and economic policy developments reported in the New York Times were associated with a percentage change in the Standard & Poor's 500-Stock Index. While the New York Times or Times front page are good indicators of stock moves, the top scientific journals like Nature and Science are not. What follows is the conclusion that if scientific breakthroughs get first reported in scientific outlets, one may invest in stock of a company holding the copyright or trademark for the innovation and simply wait until the good news hit popular media [30]. This may be the result of limited cognitive capacities, time constraints, and focused information consumption. In addition, the time of good news release plays a role [9].

5.7 Availability Biases

Availability biases may be the underlying cause of peoples' tendency to overestimate the value of their home. People overestimate the value of what they are familiar with and therefore are also prone to trade excessively in potentially suboptimal local stocks [7, 15].

Households hold underdiversified stock portfolios concentrated in firms headquartered near the city where they reside. Explanations for this local-bias assign a causal role for proximity, be it in generating an informational advantage or a familiarity bias [1].

Building on the availability heuristic, excessive media coverage may help to explain extraordinary levels of trading volume in stocks and their elevated prices of so-called overpriced glamor stocks [29]. An example of overpriced glamor stocks is given in the Internet bubble period from 1998 to 2000 but also the current cryptocurrency hype. Overpriced stocks also occur for firms that have local advantages of low competition and risk exposure [22]. Cryptocurrencies and especially social online media promoted e-currencies are believed to be another form of glamor stocks, whose market price has become detached from the actual fundamental value.

The availability heuristic of major corporations being overrepresented in the news leads to declining importance of news for smaller firms [23]. The underlying mechanism is that information only gradually diffuses among the investment public, and this is especially the case for smaller firms [23].

5.8 Quality of Information

When it comes to information as predictor of future market performance, open interests as the total number of outstanding contracts that are held by market participants at the end of the day were found to be more accurate predictors than actual future prices in the presence of hedging demand and limited risk absorption capacity in future markets [32]. Open interests comprise the total number of futures contracts or option contracts that have not yet been exercised (squared off), expired, or fulfilled by delivery. Open interest is highly pro-cyclical and correlated with macroeconomic activity and movements in asset prices [32].

5.9 Good News Breeding Overconfidence

Classic speculative bubbles are loud—price is high and so are price volatility and share turnover. Yet be aware, credit bubbles tend to be quiet—price is high but price volatility and share turnover are low as debt is less sensitive to disagreement about asset value than equity and hence has a smaller resale option and lower price volatility and turnover [26].

The addition to one of the major stock indices is associated with a price jump for the respective stock, while the exclusion from an index is likely to be followed by a price decline [3].

5.10 Bad News

In general, managerial outsourcing lowers the performance and incentives of mutual funds by about 50 basis points per year. Fund families outsource the management of a large fraction of their funds to advisory firms. After instrumenting for a fund's outsourcing status, the estimate of underperformance is three times larger. The reason for this effect may lie in the fact that an outsourced fund faces higher powered incentives; they are more likely to be closed after poor performance and excessive risk-taking [5].

References

1. Branikas, I., Hong, H. G., & Xu, J. (2016). *Location choice, portfolio choice*. Princeton University Working Paper.
2. Chang, B., & Hong, G. (2016). Assignment of stock market coverage. In *27th Annual Conference on Financial Economics and Accounting Paper*.
3. Chang, Y-Ch., Hong, H., & Liskovich, I. (2014). Regression discontinuity and the price effects of stock market indexing. *Review of Financial Studies, 28*(1), 212–246. https://doi.org/10.1093/rfs/hhu041
4. Chang, Y-Ch., Hong, H. G., Tiedens, L., Wang, N., & Zhao, B. (2015). *Does diversity lead to diverse opinions? Evidence from languages and stock markets*. Rock Center for Corporate Governance, Stanford University.
5. Chen, J., Hong, H., Jiang, W., & Kubik, J. D. (2013). Outsourcing mutual fund management: Firm boundaries, incentives and performance. *The Journal of Financial Economics, 103*(3), 454–470.
6. Chen, J., Hong, H., & Stein, J. C. (2002). Breadth of ownership and stock returns. *Journal of Finance, 66*(2), 171–205.
7. Choi, H.-S., Hong, H. G., & Scheinkman, J. A. (2014). Speculating on home improvements. *Journal of Financial Economics, 111*(3), 609–624. https://doi.org/10.1016/j.jfineco.2013.11.011
8. Cutler, D. M., Poterba, J. M., & Summers, L. H. (1988). *What moves stock prices?* National Bureau of Economic Research Working Paper 2538.
9. DellaVigna, S., & Pollet, J. (2006). *Investor inattention, firm reaction, and Friday earnings announcements*. University of California at Berkeley Working Paper.
10. Fong, K., Hong, H., Kacperczyk, M., & Kubik, J. D. (2014). *Do security analysts discipline credit rating agencies?*. Princeton University Working Paper.
11. Frankel, J., & Meese, R. (1987). Are exchange rates excessively variable? In S. Fischer (Ed.),*NBER Macroeconomics Annual* (pp. 117–152). MIT Press.
12. French, K., & Roll, R. (1986). Stock return variances: The arrival of information and the reaction of traders. *Journal of Financial Economics, 17*, 5–26. https://doi.org/10.1016/0304-405X(86)90004-8
13. Hong, D., Hong, H. G., & Ungureanu, A. (2012). *An epidemiological approach to opinion and price-volume dynamics*. American Economic Association Meeting, Chicago, January 2012.
14. Hong, H. G., & Huang, M. (2005). Talking up liquidity: Insider trading and investor relations. *Journal of Financial Intermediation, 14*, 1–13. https://doi.org/10.1016/j.jfi.2003.08.004
15. Hong, H. G., Jiang, W., Wang, N., & Zhao, B. (2014). Trading for status. *Review of Financial Studies, 27*(11), 3171–3212. https://doi.org/10.1093/rfs/hhu048
16. Hong, H. G., & Kostovetsky, L. (2010). Red and blue investing: Values and finance. *Journal of Financial Economics, 103*(1), 1–19. https://doi.org/10.1016/j.jfineco.2011.01.006
17. Hong, H. G., & Kubik, J. D. (2003). Analyzing the analysts: Career concerns and biased earning forecasts. *The Journal of Finance, 58*(1), 313–351. https://doi.org/10.1111/1540-6261.00526
18. Hong, H. G., Kubik, J. D., & Fishman, T. (2011). Do arbitrageurs amplify economic shocks? *Journal of Financial Economics, 103*(3), 454–470. https://doi.org/10.1016/j.jfineco.2011.10.007
19. Hong, H. G., Kubik, J. D., & Solomon, A. (2000). Security analysts' career concerns and herding of earnings forecasts. *RAND Journal of Economics, 31*(1), 121–144. https://doi.org/10.2307/2601032
20. Hong, H., Kubik, J. D., & Stein, J. C. (2004). Social interaction and stock market participation. *The Journal of Finance, 59*(1), 137–163. https://doi.org/10.1111/j.1540-6261.2004.00629.x
21. Hong, H. G., Kubik, J. D., & Stein, J. C. (2005). Thy neighbor's portfolio: Word-of-mouth effects in the holdings and trades of money managers. *The Journal of Finance, 60*(6), 2801–2824. https://doi.org/10.1111/j.1540-6261.2005.00817.x

22. Hong, H. G., Kubik, J. D., & Stein, J. C. (2008). The only game in town: Stock-price consequences of local bias. *Journal of Financial Economics, 90*(1), 20–37. https://doi.org/10.1016/j.jfineco.2007.11.006
23. Hong, H. G., Lim, T., & Stein, J. C. (2000). Bad news travels slowly: Size, analyst coverage, and the profitability of momentum strategies. *The Journal of Finance, 55*(1), 265–295. https://doi.org/10.1111/0022-1082.00206
24. Hong, H. G., & Rady, S. (2002). Strategic trading and learning about liquidity. *Journal of Financial Markets, 5,* 419–450. https://doi.org/10.1016/S1386-4181(02)00020-4
25. Hong, H. G., Scheinkman, J., & Xiong, W. (2005). Asset float and speculative bubbles. *The Journal of Finance, 61*(3), 1073–1117. https://doi.org/10.1111/j.1540-6261.2006.00867.x
26. Hong, H. G., & Sraer, D. (2011). Quiet bubbles. *Journal of Financial Economics, 110*(3), 503–552.
27. Hong, H. G., & Stein, J. C. (1999). A unified theory of underreaction, momentum trading, and overreaction in asset markets. *The Journal of Finance, 54*(6), 2143–2184. https://doi.org/10.1111/0022-1082.00184
28. Hong, H. G., & Stein, J. C. (2003). Differences of opinion, short-sales constraints, market crashes. *Review of Financial Studies, 16,* 487–525. https://doi.org/10.1093/rfs/hhg006
29. Hong, H. G., & Stein, J. C. (2007). Disagreement and the stock market. *Journal of Economic Perspectives, 21,* 109–128. https://doi.org/10.1257/jep.21.2.109
30. Hong, H. G., Torous, W., & Valkanov, R. (2005). Do industries lead stock markets? *Journal of Financial Economics, 83*(2), 367–396. https://doi.org/10.1016/j.jfineco.2005.09.010
31. Hong, H. G., & Wang, J. (2000). Trading and returns under periodic market closures. *The Journal of Finance, 55*(1), 297–354. https://doi.org/10.1111/0022-1082.00207
32. Hong, H. G., & Yogo, M. (2012). What does futures market interest tell us about the macroeconomy and asset prices? *Journal of Financial Economics, 105*(3), 473–490. https://doi.org/10.1016/j.jfineco.2012.04.005
33. Hong, H. G., & Yu, J. (2007). Gone fishin': Seasonality in trading activities and asset prices. *The Journal of Finance, 62*(3), 1207–1242. https://doi.org/10.1111/j.1540-6261.2007.01234.x
34. Hong, H. G., & Xu, J. (2015). *Inferring latent social networks from stock holdings.* American Finance Association 2015 Boston Meetings Paper.
35. Neiderhoffer, V. (1971). The analysis of world events and stock prices. *Journal of Business, 44*(4), 193–219. https://doi.org/10.1086/295352
36. Roll, R. (1984). Orange juice and weather. *American Economic Review, 74,* 861–880.
37. Scharfenstein, D. S., & Stein, J. C. (1990). Herd behavior and investment. *American Economic Review, 80,* 465–479.
38. Shiller, R. (1981). Do stock prices move too much to be justified by subsequent dividends? *American Economic Review, 71,* 421–436.
39. Thaler, R. H. (1987a). Anomalies: The January effect. *Journal of Economic Perspectives, 1,* 197–201. https://doi.org/10.1257/jep.1.1.197
40. Thaler, R. H. (1987b). Anomalies: Weekend, holiday, turn of the month and intraday effects. *Journal of Economic Perspectives, 1,* 169–178. https://doi.org/10.1257/jep.1.2.169
41. Zwiebel, J. (1995). Corporate conservatism and relative compensation. *Journal of Political Economy, 103,* 1–25. https://doi.org/10.1086/261973

Part IV

The Future of Behavioral Economics

Artificial Intelligence and Nudging 6

6.1 Artificial Intelligence Market Disruption

The introduction of artificial intelligence (AI) in our contemporary society imposes historically unique challenges for humankind. The emerging autonomy of AI holds unique potentials of eternal life of robots, AI, and algorithms alongside unprecedented economic superiority, data storage, and computational advantages. Yet to this day, it remains unclear what impact AI taking over the workforce will have on economic growth.

The contemporary trend of slowbalisation captures the slowing down of conventional globalization of goods, services, and foreign direct investments (FDIs) flows; yet at the same time, we still see human migration and air travel as well as data transfer continuing to rise [114]. These market trends of conventional globalization slowing and rising AI-related industries are proposed as first-market disruption in the wake of the large-scale entrance of AI into our contemporary economy. Growth in the artificial age is then proposed to be measured based on two AI entrance proxies of *Global Connectivity Index* and the *State of the Mobile Internet Connectivity 2018 Index*, which is found to be highly significantly positively correlated with the total inflow of migrants and FDI inflow—serving as evidence that the still globalizing rising industries in the age of slowbalisation are connected to AI. Both indices are positively correlated with GDP output in cross-sectional studies over the world. In order to clarify if the found effect is a sign of industrialization, time series of worldwide data reveals that Internet connectivity around the world is associated with lower economic growth from around 2000 on until 2017. A regression plotting Internet connectivity and GDP per capita as independent variables to explain the dependent variable GDP growth outlines that the effect for AI is a significant determinant of negative GDP growth prospects for the years from 2000 until 2017. A panel regression plotting GDP per capita and Internet connectivity from the year 2000 to explain economic growth consolidates the finding that AI Internet connectivity is a significant determinant of negative growth over time

J. Puaschunder, *Advances in Behavioral Economics and Finance Leadership*, Contributions to Economics, https://doi.org/10.1007/978-3-031-15710-3_6

for 161 countries of the world. Internet connectivity is associated with economic growth decline, whereas GDP per capita has no significant relation with GDP growth. To cross-validate, both findings hold for two different global connectivity measurements. The monograph then discusses a theoretical argument of dividing labor components into fluid, hence more flexible (e.g., AI), and more clay, hence more inflexible (e.g., human labor), components. These findings call for revising growth theories and integrating AI components into growth theory. AI entrance into economic markets is modeled into the standard neoclassical growth theory by creating a novel index for representing growth in the artificial age comprised of GDP per capita and AI entrance measured by the proxy of Internet access percent per country. Maps reveal the parts of the world that feature high GDP per capita and AI connectivity. The discussion closes with a future outlook on the law and economics of AI entrance into our contemporary economies and society in order to aid a successful and humane introduction of AI into our world.

Already now, about 28% of the workforce in modern economies is estimated to be based on AI or AI-supported. First-market disruptions of AI entering economies are currently speculated to cause a trend of slowbalisation—as a countertrend to globalization. Globalization sprang from America's sponsorship of a new world order in 1945, which allowed cross-border flows of goods and capital to recover after years of war and chaos [27–29]. During the golden age of globalization from 1990 to 2010, the world became flat: Immigration increased from 2.9 to 3.3% of the world's population and global trade grew from 39% of GDP in 1990 to 58% last year [158]. Asia became part of the globalized upon China's entry into the WTO in 2001, which created a model of offshoring manufacturing to countries based on cost efficiency variances, primarily labor costs [108]. The Washington Consensus embraced the world and promised to bring prosperity to everyone around the globe [145]. Open markets and free trade were praised to lift billions of people out of poverty in Asia, Latin America, and Africa via economic growth [61].

With the collapse of the Soviet Union in 1989 and the end of the Cold War in 1991, the world became even more interconnected and global market economies integrated around the world. Trade and investment increased, while barriers to migration and cultural exchange lowered [90]. The European Union but also free trade agreements, such as the North American Free Trade Agreement (NAFTA), which the governments of the United States, Canada, and Mexico signed in 1992, removed barriers to the free flow of people, goods, and services, thereby facilitating greater trade, investment, and migration across borders in an unprecedented way [108, 112, 171].

During the last 20 years, China increased its GDP significantly, a sign of historically unprecedented growth for a country of this size [108]. A similar phenomenon occurred in India, Vietnam, and other countries. Globalization also supported the growth of large multinational companies that offshored production processes and consumers to access an endless number of products at competitive prices from around the globe. Commerce soared as the cost of shifting goods in ships and planes fell, phone calls got cheaper, tariffs were cut, and finance liberalized. Business went gangbusters as firms set up around the world, investors roamed, and consumers shopped in supermarkets with goods from around the globe [108]. As never before in history, traveling had become available to the general populace at affordable prices. The number of refugees reached all-time highs. If not moving oneself, free data services provided on the "window to the world" Internet, allowed everyone to consume the globe anytime anywhere.

Yet, globalization also brought about negative consequences and unforeseen shadows of the invisible hand. Until the 1990s, studies report no connection between GDP and happiness—yet from the 1990s, on there is a negative correlation found between GDP and happiness [70]. This trend is attributed to the Internet and access to information about other places on earth's living conditions creating emotionally hurtful comparisons in desolate places, also fueling migration trends, which have never been higher than as of now.

6.1.1 Slowbalisation

When America took a protectionist turn in its 2016 Presidential election they were, once again, first in sensing and acting on a contemporary detected, most novel worldwide trend: We currently live in the age of slowbalisation. Protectionism, trade wars, emerging economies' slowdown, and the decrease in goods and services trade as well as a slump in transnational investments are all signs of the global trend of globalization that have come to a halt. United Kingdom followed shortly after the U.S. presidential with voting for Brexit. Globalization has slowed in our current times of "slowbalisation," a term coined in 2015 by Adjiedj Bakas, who sensed first that globalization has given way to a new era of sluggishness. The 2022 unfolding Ukraine crisis is another geopolitical event that drives the trend towards protectionism with the World Trade Organization noticing a long-term rewind of globalized world economies.

Globalization has slowed in the past decade after the 2008 global recession. Trade has fallen from 61% of world GDP in 2008 to 58% prior to COVID-19 [158]. If these figures exclude emerging markets (of which China is one), it has been flat at about 60% [158]. The capacity of supply chains that ship half-finished goods across borders has shrunk. Intermediate imports rose fast in the 20 years to 2008, but since then have dropped from 19% of world GDP to 17% by 2019 [158]. The march of multinational firms has halted as the global corporate share of global profits of all listed firms has dropped from 33% in 2008 to 31% by 2019 [158]. Long-term cross-border investment by all firms, known as foreign direct investment (FDI), has tumbled from 3.5% of world GDP in 2007 to 1.3% in 2018 [158]. As cross-border trade and companies have stagnated relative to the economy, so too has the intensity of financial links. Cross-border bank loans have collapsed from 60% of GDP in 2006 to about 36% [158]. Excluding rickety European banks, they have been flat at 17%. Gross capital flows have fallen from a peak of 7% in early 2007 to 1.5% [158]. Since 2008, the share of economies converging from emerging economies to catch up with the rich world in terms of output per person using purchasing power parity has fallen from 88 to 50% in 2019 [158]. So in fact, almost all conventional measures of global trade and market integration have fallen. Tariffs have reached the highest levels in the last 40 years, and additional costs of trade have begun to be passed onto consumers [108]. In the second half of 2018, the largest US companies lost about 6 billion—or 3%—in profits due to tariffs [108, 158]. US and Chinese investments in Europe have fallen dramatically, for instance, China's investment by 73% in 2018 [158]. The global value of foreign investment by multinationals decreased by 20% in the same year [158]. As the service sector appears to continue to expand, relocation for the sake of consumption has stagnated or declined as it is harder to relocate services [25, 39]. Based on the last decade, *The Economist* [158] predicts a decline in exports from 28 to 23% of GDP over the next 10 years, which would resemble a similar drop between 1929 and 1946.

Slowbalisation speaks to the fact that since the 2008 World Financial Recession, Asia's growth rates are slowing, cross-border investments, trade, bank loans, and supply chains have been shrinking or stagnating relative to world GDP [158]. While one of the main benefits of globalization was that between 1990 and 2010, most emerging countries were able to close some of the gaps with developed ones, and a slowdown in globalization likely leads to a reversal in underdeveloped parts of the world catching up [158]. In addition to projected major political risks and the decline in socio-economic development, with the absence of a global cooperation, it will be more difficult to tackle and solve major coordination challenges such as climate change and climate refugees, immigration, and tax evasion [10, 108, 115]. This predicament is crucial if we seem to trade-off environmental degradation with international development opportunities—the two most pressing obstacles for contemporary humankind [172].

Politically, where we seemed to have spent decades after two world wars to break down walls and pacify Europe in a Union, we are now back to building barriers faster than before [108]. Since 2009, the number of new free trade agreements between countries has plummeted and restrictions on trade have proliferated on duties, anti-dumping measures, and Non-Tariff Barriers to trade (NTBs). Bloomberg reports that the DHL monitor tracking shows that global trade is continuing to lose a little steam amid an escalating tariff battle between the world's biggest economies [108].[1] Media and news but also big data trends appear to have open gates to the world as never before while shrinking the number of local newspapers and media outlets [58]. Corporate greed and politics of fear are partially argued as socio-political trends around slowbalisation [108]. International remedies—such as science diplomacy—are called upon to ensure upholding the

[1] www.bloomberg.com/news/articles/2018-09-27/global-trade-growth-slowly-losing-steam-as-business-feels-pinch.

benefits of globalization in our commonly shared fragile world to ensure continuous economic prosperity, societal advancement, and humane dignity for all [13].

Yet, this is not the end of the story, as some globalization features still show rising integration. Technological advances, including mobile phones and especially the Internet, have contributed to globalization by connecting people all over the globe. Innovation spurs companies to substitute labor, while technology shocks drive economic growth, especially when technologies progressively reduce the physical work component [170]. While goods are not shipped around the globe in extensive global value chains, the consumers themselves have become yet more global. The World Wide Web links billions of people and devices, providing innumerable opportunities for the exchange of goods, services, cultural products, knowledge, and ideas. The Internet connectivity and volume of data crossing borders have risen by 64 times; according to McKinsey, people appear to enjoy experiences abroad and consume data.

Building dreams and hope based on information shared online, migration to the rich world has risen over the past decade. International parcels and flights are growing fast, almost exponentially.

Migration has also increased sharply in the last ten years. International transfers and air traffic grew exponentially in the period before the Coronavirus crisis. Until shortly before the onset of the COVID-19 pandemic, air traffic was the highest ever.

This trend of polarization between ongoing polarization of globalization on data and people versus slowbalisation of traditional goods and services as well as finance is argued as the first sign of AI entering economic production and changing goods and service trade. Technological and political factors could indicate a market disruption that has already begun and currently echoes in globalization versus slowbalisation occurring parallel to each other. The currently described trend of slowbalization could only be a forerunner of the upcoming digitization disruption, which will create a whole new world in the long term due to the COVID-19 pandemic, which is fundamentally different from the one we know.

In the decade prior to COVID-19, an already ongoing digitalization disruption heralded as big data allowed a set of innovative firms, including social media, online commercial platforms, and search engines to reap skyrocketing profits that often remain untaxed. These economic gains are concentrated in areas such as big data hoarding, the sale of behavioral data about consumers, and targeting online audiences with customized advertisement [109, 111].

From an economic perspective, COVID-19 is an external shock that has accelerated ongoing digitalization trends [126]. COVID-19 has perpetuated the online tech world. Physically distant, we became closer digitally than ever before. Worldwide data traffic exploded on a flat digitalized globe. Because of widespread lockdowns, "social distancing" and increased home office work in many industries, social scientists have observed a more widespread acceptance for instant communication tools, social engagement, and entertainment platforms [32]. We can thus say that certain firms and industries have benefited from the pandemic while many others have suffered from the expenses and burdens of COVID-19 [6, 7, 36, 75, 159]. Traditional small businesses appear to be particularly vulnerable [17, 37, 76, 105].

Today's economies around the world are affected by the pandemic and health crisis. Future innovations in the medical field are predestined to be in the realm of digitized health care and self-monitoring with the help of electronic self-measuring devices and modern molecular genetic analyzes. These can make the use of new active ingredients in the field of prevention of serious diseases more efficient. In post-COVID economies, digitalized hygiene and healthcare are likely to be further advanced as healthier workers around the world will have a competitive advantage [161].

Employers may be more interested in what category their workforce may fall to plan workplace safety precaution measures when building healthy working conditions [40, 41]. More than ever before in the history of modern industrialization do employers nowadays care about the overall well-being and physical interaction of their labor cadre in a hygienic environment. Respective preventive medical care of the workforce and community-building around monitoring of one's own and others' health but also group learning how to enhance hygiene in teams will gain more attention in the COVID-19-struck workplace and will have lasting changes enacted [40, 41].

In light of social distancing mandates and with the growth of scientific evidence derived from algorithms and big data, workers with better access to internet connectivity and AI-human-compatibility (i.e., computer and AI literacy, and related skills) have growing competitive advantages [40, 41]. It may also become a matter of survival for large organizations to understand the health of their workers with the help of these novel technologies. On the one hand, employers will need to estimate whether workplace conditions are likely to produce mass outbreaks—such as the ones that, for instance, occurred in the meatpacking industry or luxury tourism cruise ships in several countries [42, 47]. On the other hand, and maybe less benignly, employers will want to know whether workers' medical histories, genetic profiles, living arrangements and social habits are likely to result in COVID-19 infection risks and predict trajectory likelihoods based on genetic prevalence derived from big data analyses.

With the entry of AI algorithms and insights derived from large data sets in the medical field, they may hope to maintain a healthy workforce through encouraging workers' self-monitoring, while also pro-actively caring for safety through mobile tracking of infected as a means of crowd control, as well as potentially through the use of predictive algorithms [53]. Much like many employers require drug tests at the hiring stage or periodically, they may seek to use digitalization tools to predict health statuses and working conditions' outcomes to reduce the risk of being put out of business or severely harmed by a COVID-19 outbreak. Arguably, firms that are better able to use technology to determine and track employees' health status will be the winners in the post-pandemic market. It may be too early to say whether nimble startups and other small- and medium-sized firms will be more likely to succeed, or whether large corporate behemoths with access to large datasets will be more likely to thrive [54]. Already, however, it is becoming apparent that these novel digitalization opportunities come with the price of a heightened responsibility to protect privacy in retrieving big data inference, ensure access to information and healthcare democratically, secure individuals from discrimination against health status propensities, and back those who are naturally hindered to compete in markets financially and socially [137].

In light of the current political East-West tensions and bottleneck shortage problems around the world, another trend got also exacerbated. Reshoring appears to occur, in which domestic technology-enhanced production is favored over outsourcing to desolate low-skilled, low-income territories. With the ringing in AI revolution, technological development is bringing production and manufacturing closer to the end user. Fourth Industrial Revolution robots are expected to become more efficient and affordable. With that, conventional globalization practices—such as offshoring manufacturing to cheap labor cost countries—will most likely decline. Reshoring will bring back production to where goods and services are actually and finally consumed. The most obvious example is energy and a prospective attempt to decentralize renewable energy generation. Your solar panel becomes more productive if energy needs not to be stored but simply can be shared with your neighbor when not needed.

AI holds the potential to replicate human existence but live eternally. 24/7 working robots that can live eternally are expected to become the driver of industrialized economies and replace the majority of human workforce [79]. 3D-printing techniques and nanotechnology that allow production to start at the molecular or even atomic level are fostering reshoring as relocating production sights from global value chain sights that were spread out during the golden years of globalization to where goods and services are consumed today.

Reshoring of global production closer to where consumers appear favorable in light of climate change and carbon emissions; yet shunning low-skilled labor in developing parts of the world from production for globally operating multinationals may revert international development [12, 57, 91–93]. So while companies around the globe featured an offshoring trend during the golden age of globalization, contemporary reshoring and glocalization occurs.[2] Slowbalisation appears to strengthen regional trade blocs, especially in Europe and Asia [108, 158]. Corporations tend to be focusing their production back to where they serve their customers and consumers have recently gained substantial interest in more local products.

There is a projected impact of robotic development on international trade. Robots are expected to be more accurate and work 24/7, while being less demanding than human workers. Millions of employees may lose their jobs over the next few decades, substituted by robots. In addition, advances in 3D printers may soon make it possible to substitute large factories with much smaller ones, closer to the consumer, where the manufacturing process is simplified thanks to the reproduction of models [2]. New materials could be manufactured near the consumer via

[2] https://en.wikipedia.org/wiki/Glocalization.

3D printing and/or 3D-printing drones, in order to substitute natural materials that need to be transported from distant mines and deposits [164]. Trade links within regional blocs may increase and blocks become more homogenous, both in Europe and Asia.

High-end production has discovered the luxury of opening consumers' eyes for the entire production and ensuring that corporate social responsibility is lived throughout the value change. Moreover, when companies bring production back into their countries for AI, unskilled workers lose out in the domestic markets while leaving behind markets that flourished due to outsourcing companies.[3] Reshoring means that former outsourced tasks are simply performed by AI in high-skilled interconnected countries, with whom low-skilled workers in the developing world will have to compete. The transition to the *new globalization* has caused the workers in developed markets to lose bargaining power as they now operate in the production phases that are most vulnerable to delocalization and automation, while the Western world will face competition with AI in wage-stagnating economies [10, 16, 108].

A trend which will—for instance—pit a 5G automated device pit against a low-skilled worker in a desolate place on earth with not even Internet access, which allows learning and productivity gains but also drives migration and brain drain [80]. Slowbalisation and reshoring are thereby expected to widen the gap between the rich and the poor. AI entering our economies may lead to a trend of reshoring, and thereby shunning away international low-cost production sights from global production. The global gap between AI-automated hubs and non-automated places on earth will therefore likely increase in the years to come. So while reshoring offers opportunities for more sustainable production in light of climate change, when we consider the environmental impacts of shipping goods around the globe until they reach the end user, in the end, it also bears the risk of restricting global economic development.

The 2019 coronavirus crisis has taken all these digitization trends and slowbalization to a new level. From an economic point of view, the lockdowns have strengthened tendencies against globalization that existed before the crisis. Already in the decade before COVID-19, globalization slowed in global trade in goods, finance and profits abroad. Global production value chains were brought closer to consumers again in the course of the so-called glocalization, reshoring and slowbalization trends. Global interconnectedness risks have been foreseen in the financial sector, food supply, and global healthcare [30]. Nationalism increased in homeland-focused politics and EU exit sentiment [52, 108].

The new Coronavirus crisis has now amplified these trends and, above all, put an end to public mobility and international tourism traffic. The clear losers from the COVID-19 pandemic are tourism and industries around the world. Across finance, the FTSE, Dow Jones Industrial Average and Nikkei all fell sharply and in their largest quarterly decline since 1987, but only for a short period of time [67]. Just

[3] https://www.cgdev.org/publication/middle-class-winners-or-losers-globalized-world.

six months after the slump in economic output at the beginning of 2020, the Standards and Poor's market index in the USA was back at the level it was before the pandemic. In view financial market recovery, the real economy is still being pushed to the limits of what is possible. Unemployment has risen rapidly since the outbreak of the crisis. At the end of June 2020, the U.S. unemployment rate had risen from 3.7 to 10.4%, in Great Britain from 3.8 to 4.8% and less strongly in Germany—due to short-time work models—from 3.2 to 3.9% [62].

As of mid-June 2020, as a result of lockdowns and fears about shopping in stores, annual percentage changes in consumption in China fell by 32%, in Germany by 49%, in the USA by 50% and in Great Britain by as much as 78%. The OECD estimates that 19% of workers in the UK were furloughed, 23% in Germany and as much as 41% in France. The International Monetary Fund assumes that the majority of countries already have a 5–15% chance of recession based on economic data for the first quarter of 2020 and as measured by total Gross Domestic Product change. The global economy has contracted by 3% overall and is estimated to have contracted by 3–5% in 2020.

At the same time, the Coronavirus crisis perpetuated digitization and led to unprecedented applications of digital communication. During the lockdowns, there has been an abrupt digital disruption, considering online communication, learning, working, consuming and medical consultations. The digitization disruption grew qualitatively but also quantitatively exponentially during the outbreak of the coronavirus pandemic. The 5G revolution that is currently taking place should now reinforce all of these trends. The major downside of the digitalization disruption is the reported exponential increase in online crime around the world based on the State of the Internet Report.[4]

6.2 Macroeconomic Modeling

Globalization led to an intricate set of interactive relationships between individuals, organizations, and states [27, 28]. Unprecedented global interaction possibilities have made communication more complex than ever before in history as the whole has different properties than the sum of its increasingly diversified parts [29]. Electronic outsourcing in the age of artificial intelligence is likely to increase and with this trend a possible societal divide in the twenty-first century [109]. The AI revolution appears to be different from conventional technology shocks as the electronic information share and big data generation open novel and yet unregulated opportunities to reap surplus value from social media consumers [109]. For one, social media space can be sold to marketers who can constantly penetrate the consumer-worker in a subliminal way with advertisements. But also nudging occurs as the big data compiled about the social media consumer-worker can be

[4] https://www.akamai.com/our-thinking/the-state-of-the-internet/global-state-of-the-internet-security-ddos-attack-reports.

resold to marketers and technocrats to draw inferences about consumer choices, contemporary market trends or individual personality cues used for governance control, such as border protection and tax compliance purposes [109].

Addressing these novel economic growth components in the nudgital society allows us to better govern value creation in the digital age, leading to the potentially unequal accumulation and concentration of power following the greater goal to improve capitalism and democracy in the digital artificial age [110, 115]. In light of growing tendencies of globalization, the demand for an in-depth understanding of how information will be shared around the globe and artificial intelligence hubs may evolve in economically more developed parts of the world has gained unprecedented momentum [14, 74]. In addition, robotics and AI self-learning algorithms appear to resemble more human features than conventional technologies. The legal status also differs from AI being assumed to be quasi-human. First robots have gained citizenship and the legal codification of AI in common law countries bestows robots quasi-human legal status and applies the civil code in the writing of legal codification to guide on the AI introduction in markets and our contemporary society.[5]

Unprecedented value opportunities from information sharing and AI being considered quasi-human, economic growth in the artificial age may be different than neoclassical growth theory would suggest. If considering that AI takes over traditional labor and leads to a reduction of conventional production, conventional growth in the artificial age may decline. Reaping value from unconventional new AI productivity may not be captured in standard neoclassical growth components of conventional capital and labor—as AI's relation to capital and labor is unclarified. AI on the one hand seems to resemble or being treated as quasi-human but on the other hand is very different from labor as for the eternal living capacities and computation power as well as interchangeability.

The artificial intelligence revolution will expand our concept of time as artificial intelligence has eternal life and 24/7 productivity capacities will change tact and lifespan depreciation rates. Algorithms improving behavioral decision-making biases are also not covered in capital and labor output [18–21]. Productivity of the sharing economy or reaping value from big data may not be displayed in standard growth components, and AI is neither capital nor labor. Sharing information over a mobile application is also neither capital nor labor.

Potential effects of AI on economic growth are a replacement of labor with capital as Aghion et al. argue in 2017 based on evidence from the field.[6] Yet to this day, there is no clear empirical investigation of AI's impact on ordinary production of goods and services with potential effects on growth rates and income shares [2]. While AI may help solve complex problems and save on computation time, the data computation storage may not be integrated or reflected by standard growth theories, for sure not in exogenous but also not so much in endogenous

[5] https://techcrunch.com/2017/10/26/saudi-arabia-robot-citizen-sophia/.

[6] https://scholar.harvard.edu/files/aghion/files/artificial_intelligence.pdf.

growth theory versions. Endogenous growth theory may address learning opportunities but may not accurately cope with the novel data storage and computational advantages of AI, which may increase the scope of new production lines while driving trends of reshoring and bringing back production closer to where the design and planning occurs. Reshoring may impact finance and human capital flows [26, 141, 163]. Human substitution through AI—such as inventing new ideas and new creative technologies—may not be captured properly in contemporary growth theories as well [80]. AI may become rapidly self-improving and should be seen as a producing singularity that features unbounded machine learning intelligence and economic growth eternally [2]. Aghion et al. [2] put forward the first integration of AI as a separate component within growth theory, so neither capital nor labor. All these features of AI encroaching markets demand revising growth theories in light of a potential currently ongoing AI market disruption in order to draw inferences on how to revise growth theory in the artificial age.

In order to clarify if the currently detected slowbalisation trend is the first sign of a market disruption related to AI entering markets, the empirical investigation features Study 1 to (1) show that the currently detected polarization of globalization and slowbalisation trends is AI market introduction driven. Study 1 validates the slowbalisation trend with particular focus on proving evidence for still ongoing globalization being connected to AI-led growth.

As for the introduction of AI into contemporary economic markets, the empirical part will then (2) estimate country differences in economic output and AI infiltration of the market in cross-sectional between-country studies. Study 2 aims to clarify if AI is positively associated with economic output. Study 2 therefore captures GDP in the artificial age in a cross-sectional comparison between countries and offers a cross-validation check operationalized by two different AI market entrance proxies, the *Global Connectivity Index*[7] and the *State of the Mobile Internet Connectivity 2018 Index*.[8]

In order to shed light on the relation of GDP growth and AI, Study 3 will (3) estimate economic growth and AI over time in cross-validated time series as well as a panel regression. Study 3 outlines the relation of AI entering markets and economic growth in a time series of economic growth for the years 1961 until 2017. The same cross-validation check of measuring the AI entrance in markets by two different proxies is given in Study 3. In order to further validate the findings and distinguish the relation of AI and growth over time from the general relation of industrialization and technological advancement leading to lower growth rates, a cross-sectional regression is calculated for explaining GDP growth by AI entrance and GDP per capita. To further validate if GDP growth is systematically related to AI entrance, a fixed-effect panel regression is calculated for explaining the relation of GDP growth to AI entrance and GDP per capita.

[7] http://www.huawei.com/minisite/gci/en/country-rankings.html.

[8] http://www.huawei.com/minisite/gci/en/country-rankings.html.

In order to consolidate the observation that there is a slowbalisation trend in conventional globalization parameters while globalization continues in the AI domain, a correlation study will be staged. As a proxy for AI entering economic markets, Internet connectivity, as measured by the *Global Connectivity Index*[9] will be related to GDP pillars of agriculture, industry, and service sectors as derived from the World Bank dataset on GDP of the year 2017 and a cross-validation check be performed with the *State of the Mobile Internet Connectivity 2018 Index.*[10] This measure should aid in understanding what GDP sector AI is attributed to. Further, the different components of the slowbalisation trend will be related to one another in a correlation study in order to see whether slowbalisation is a sign of AI entering markets and growth theory not being able to truly capture AI productivity. A trend of globalization still continuing in AI-featuring industries and countries will be highlighted by relating AI integration with globalization hallmarks of capital and labor movements.

Conventional growth theory components may not capture AI-led growth. A macroeconomic study will investigate the relation of Internet connectivity as a proxy for AI integration into the economy and GDP. In order to operationalize and consolidate the underlying premise that a currently ongoing AI introduction into markets may not be captured appropriately by standard economic growth theories, economic growth measured by GDP[11] will be retrieved online from the World Bank database for the year 2017 and related to Internet connectivity based on the *Global Connectivity Index*[12] as an indicator of AI market potential.

As a cross-validation check, Internet connectivity for the year 2018 will be considered by the *State of the Mobile Internet Connectivity 2018 Index*[13] in order to investigate another measure of Internet connectivity in relation to economic output as measured in GDP output derived from the World Bank dataset on GDP.[14]

In order to operationalize and consolidate the underlying premise that an AI market entrance may not be fully captured by economic growth theories, economic growth measured by GDP[15] will be retrieved online from the World Bank database for the years 1961 until 2017 and Internet connectivity based on the *Global Connectivity Index*[16] homepage as an indicator of market relevance and innovation potential. Time windows of economic output and AI presence will be compared for the periods of 2000–2017, to mark the beginning of AI entering markets, and 2008–2017 to control for post-economic crises worldwide markets.

---

[9] http://www.huawei.com/minisite/gci/en/country-rankings.html.

[10] https://www.gsma.com/mobilefordevelopment/resources/state-of-mobile-internet-connectivity-2018/.

[11] https://data.worldbank.org/indicator/ny.gdp.mktp.kd.zg.

[12] http://www.huawei.com/minisite/gci/en/country-rankings.html.

[13] http://www.huawei.com/minisite/gci/en/country-rankings.html.

[14] https://data.worldbank.org/indicator/ny.gdp.mktp.kd.zg.

[15] https://data.worldbank.org/indicator/ny.gdp.mktp.kd.zg.

[16] http://www.huawei.com/minisite/gci/en/country-rankings.html.

Both time windows will be compared with the rest of the data, which spans from 1961 until 1999 for a worldwide dataset.

As a cross-validation check, Internet connectivity for the year 2018 will be considered based on the *State of the Mobile Internet Connectivity Index*[17] over time in order to investigate another measure of Internet connectivity and technology in relation to economic growth. Time windows of economic output and AI presence will be compared for the periods of 2000–2017, to mark the beginning of AI entering markets, and 2008–2017 to control for post-economic crises worldwide markets with the rest of the data, which spans from 1961 until 1999 for a worldwide dataset. Comparing time windows of pre-AI introduction and post-AI entrance in markets allows for testing a unique relation of AI and economic growth around the world and over time.

Three regression studies will capture the relation of the entrance of AI to economic growth: Regression 1 measures the relation of GDP growth with AI entrance by the proxies of *Global Connectivity Index*[18] and the *State of the Mobile Internet Connectivity 2018 Index.*[19] Regression 2 relates GDP growth with AI entrance by the proxies of *Global Connectivity Index*[20] and the *State of the Mobile Internet Connectivity 2018 Index*[21] and tests for an additional relation with GDP per capita in a cross-sectional comparison for countries of the world in 2017. Regression 3 relates GDP growth with AI entrance and cross-validates the previous findings by using the World Bank data on *Individuals Using the Internet*[22] in percent as a proxy for AI entrance and tests for an additional relation with GDP per capita in a fixed-effect panel regression over time employing different time windows capturing pre- and post-AI market entrance.

As for testing for a concurrent slowbalisation trend in relation to GDP components, the *State of the Mobile Internet Connectivity 2018 Index*[23] served as a proxy for AI entrance in markets and was found to be highly significantly positively correlated with the service sector percentage of the entire GDP composition ($r_{\text{Pearson}} = 0.605$, $n = 161$, $p < 0.000$) and highly significantly negatively correlated with the agriculture GDP sector percentage of the entire GDP composition ($r_{\text{Pearson}} = -0.763$, $n = 161$, $p < 0.000$). The State of the Mobile Internet Connectivity 2018 Index[24] was found to be highly significantly positively correlated with the total inflow of migrants ($r_{\text{Pearson}} = 0.263$, $n = 161$, $p < 0.001$) and foreign direct investment (FDI) inflow ($r_{\text{Pearson}} = 0.298$, $n = 159$, $p < 0.000$). As a

[17] http://www.huawei.com/minisite/gci/en/country-rankings.html.
[18] http://www.huawei.com/minisite/gci/en/country-rankings.html.
[19] http://www.huawei.com/minisite/gci/en/country-rankings.html.
[20] http://www.huawei.com/minisite/gci/en/country-rankings.html.
[21] http://www.huawei.com/minisite/gci/en/country-rankings.html.
[22] https://data.worldbank.org/indicator/it.net.user.zs.
[23] https://www.gsma.com/mobilefordevelopment/resources/state-of-mobile-internet-connectivity-2018/.
[24] https://www.gsma.com/mobilefordevelopment/resources/state-of-mobile-internet-connectivity-2018/.

cross-validation check, the *State of the Mobile Internet Connectivity 2018 Index*[25] is highly significantly positively correlated with Internet connectivity as measured by the *Global Connectivity Index*[26] ($r_{\text{Pearson}} = 0.894$, $n = 79$, $p < 0.000$), which will serve as a proxy for cross-validating findings in Studies 2 and 3.

Regarding the relation of GDP and AI entrance in economic markets, GDP data was derived for the year 2017 from the World Bank database on world economic output.[27] AI penetration of markets was measured by the proxy of the *Global Connectivity Index*[28] of 79 countries. The world country's GDP and the *Global Connectivity Index* are highly significantly positively correlated ($r_{\text{Pearson}} = 0.344$, $n = 77$, $p < 0.000$).

As a cross-validation check, AI presence in markets was also measured by the proxy of the *State of the Mobile Internet Connectivity 2018 Index*[29] of 162 countries. The world country's GDP and the *State of the Mobile Internet Connectivity 2018 Index* are highly significantly positively correlated ($r_{\text{Pearson}} = 0.244$, $n = 160$, $p < 0.002$).

In order to study if the found effect is a sign of industrialization and negative GDP growth, time window studies and a regression plotting GDP per capita and Internet connectivity to explain GDP growth were staged.

Internet connectivity data of 79 countries based on the *Global Connectivity Index*[30] homepage was categorized into lowest, lower, higher, and highest Internet connectivity countries.

The lowest Internet connectivity had Ethiopia (23), Bangladesh (24), Bolivia (25), Pakistan (25), Tanzania (25), Uganda (25), Paraguay (26), Botswana (29), Ghana (29), Kenya (29), Namibia (29), Nigeria (29), Ecuador (31), Algeria (32), India (33), Indonesia (33), Morocco (33), and Venezuela (33).

Low Internet connectivity countries were Egypt (34), Jordan (34), Lebanon (34), Vietnam (34), Philippines (35), Peru (37), Argentina (38), Colombia (39), Serbia (39), Turkey (39), Thailand (40), Ukraine (41), Uruguay (41), Kazakhstan (42), Mexico (42), Oman (42), South Africa (42), Brazil (43), Belarus (44), Bulgaria (44), and Saudi Arabia (44).

High Internet connectivity featured Bahrain (45), Kuwait (45), Poland (45), Romania (45), Croatia (46), Greece (46), Russian Federation (46), Chile (48), Malaysia (48), Hungary (49), Slovak Republic (49), Czech Republic (50), Italy (50), China (51), Slovenia (51), Lithuania (52), Portugal (52), United Arab Emirates (53), Estonia (54), Spain (55), and Austria (60).

---

[25] https://www.gsma.com/mobilefordevelopment/resources/state-of-mobile-internet-connectivity-2018/.

[26] http://www.huawei.com/minisite/gci/en/country-rankings.html.

[27] https://data.worldbank.org/indicator/ny.gdp.mktp.cd.

[28] http://www.huawei.com/minisite/gci/en/country-rankings.html.

[29] https://www.gsma.com/mobilefordevelopment/resources/state-of-mobile-internet-connectivity-2018/.

[30] http://www.huawei.com/minisite/gci/en/country-rankings.html.

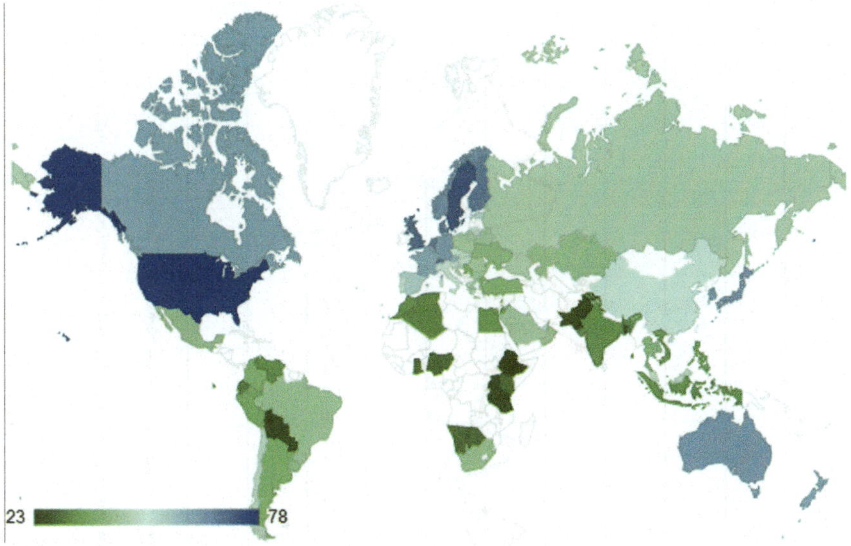

Graph 6.1 World Internet connectivity from highest (dark blue) to lowest (dark green) based on the Global Internet Connectivity index for 2017

The highest Internet connectivity had Belgium (61), France (61), Canada (62), Ireland (62), New Zealand (62), Germany (63), Luxembourg (63), Australia (64), South Korea (64), Japan (65), Norway (65), Netherlands (67), Denmark (68), Finland (68), United Kingdom (70), Switzerland (71), Sweden (73), Singapore (75), and the United States (78).

Graph 6.1 outlines the world Internet connectivity by highly Internet-connected countries in dark blue, high Internet-connected countries in blue, low Internet-connected countries in green, and lowest Internet-connected areas in dark green based on the Global Internet Connectivity index for 2017.

Graph 6.2 outlines the relation of Internet connectivity and economic growth over time. From 2000 on Internet-connected areas appear to show slower GDP growth rates than countries with the lowest Internet connectivity.

Regarding the relation between GDP and AI entrance in economic markets over time, GDP growth data was derived for the years from 1961 until 2017 from the World Bank database on world economic output.[31] AI penetration of markets was measured by the proxy of the *Global Connectivity Index*[32] of 79 countries. The world country's GDP growth over time was calculated for time compartments prior to the Internet revolution and after the Internet revolution as well as compartments for prior and after the 2008 World Financial Recession. While Internet connectivity is obviously not related to GDP growth prior to 2000, GDP growth is highly

[31] https://data.worldbank.org/indicator/NY.GDP.MKTP.KD.ZG.

[32] http://www.huawei.com/minisite/gci/en/country-rankings.html.

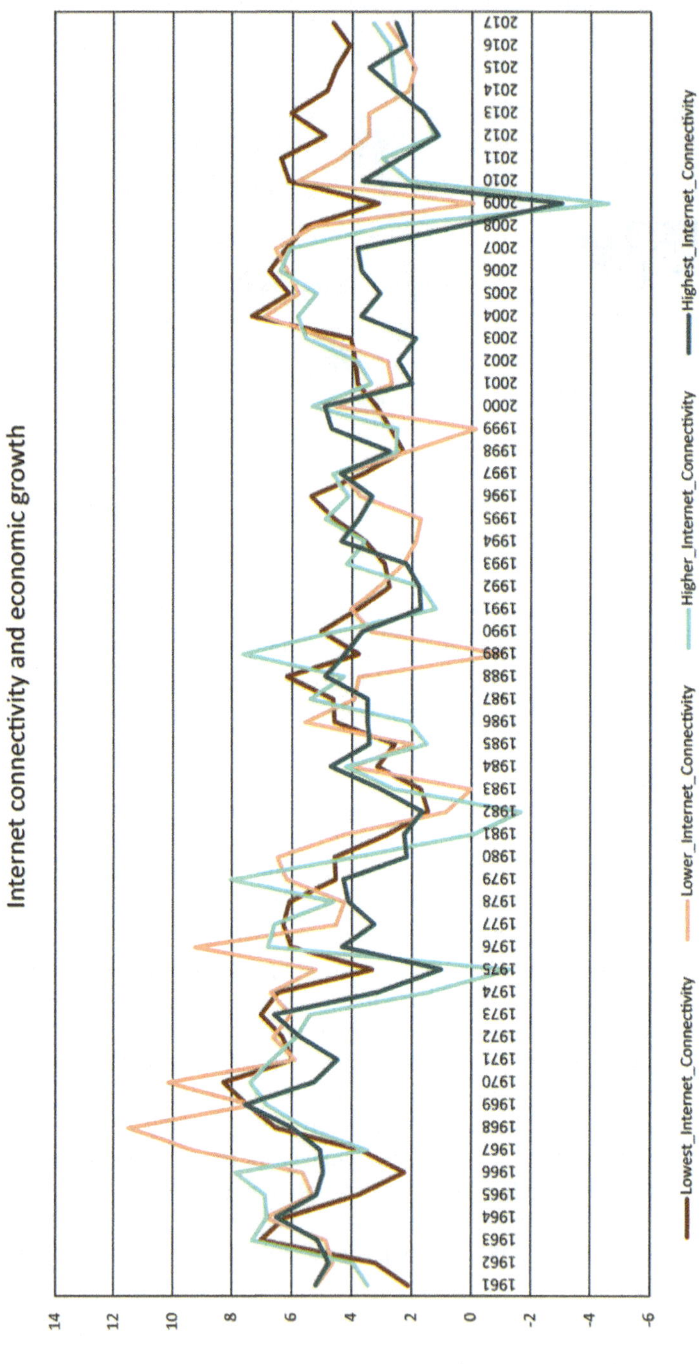

Graph 6.2 Relation of Internet connectivity based on the *Global Connectivity Index* (http://www.huawei.com/minisite/gci/en/country-rankings.html) and economic growth over time

significantly negative correlated with Internet connectivity for the period of the years 2008–2017 ($r_{\text{Pearson}} = -632$, $n = 79$, $p < 0.000$).

As a cross-validation check, Internet connectivity data of 162 countries based on the *State of the Mobile Internet Connectivity 2018 Index*[33] was categorized into lowest, lower, higher, and highest Internet connectivity countries.

The lowest Internet connectivity had Niger (18.56), Chad (18.73), Afghanistan (20.41), Malawi (23.66), Burundi (24.67), Burkina Faso (26.24), Democratic Republic of Congo (26.76), Mali (27.81), Guinea-Bissau (28.14), Gambia (30.95), Mozambique (31.03), Zambia (31.48), Togo (31.97), Madagascar (33.01), Liberia (33.08), Mauritania (33.48), Haiti (33.85), Sierra Leone (34.75), Uganda (36.49), Yemen (36.81), Pakistan (37.08), Benin (37.25), Senegal (37.3), Ethiopia (37.68), Timor-Leste (38.7), Nepal (39.11), Tanzania (39.4), Sudan (39.71), Rwanda (40.01), Zimbabwe (41.63), Republic of Congo (42.04), Cameroon (42.76), Tajikistan (43.77), Lesotho (43.99), Namibia (45.25), Lao (45.31), Cote d'Ivoire (45.73), Nigeria (45.91), Solomon Islands (45.91), and Papua New Guinea (46.03).

Low Internet connectivity countries were Uzbekistan (46.31), Iraq (46.46), Gabon (47.68), Cambodia (47.99), Bangladesh (48.35), Angola (48.84), Kenya (50.95), Botswana (51), Kyrgyzstan (51.03), Ghana (52.73), Myanmar (53.22), Azerbaijan (53.31), Bhutan (53.57), India (53.67), Armenia (54.27), Honduras (54.91), Vanuatu (55.39), Guyana (55.59), Sri Lanka (55.63), Algeria (55.93), Nicaragua (56.08), Cabo Verde (56.17), El Salvador (56.2), Egypt (56.45), Tonga (57.77), St. Lucia (57.99), Morocco (58.04), Libya (58.79), Mongolia (58.79), Iran (59.43), Belize (59.55), Bosnia and Herzegovina (59.58), South Africa (59.89), Fiji (60.13), Georgia (60.2), Tunisia (60.38), Macedonia (60.58), Guatemala (60.75), Jordan (60.84), and Indonesia (61.12).

High Internet connectivity featured Dominican Republic (61.5), Jamaica (62.23), Venezuela (62.4), Panama (62.83), Vietnam (63.03), Ukraine (63.24), Bolivia (63.57), Kazakhstan (63.58), Samoa (63.79), Costa Rica (64.27), Mauritius (64.66), Brazil (64.76), Paraguay (64.78), Colombia (64.81), Trinidad and Tobago (64.86), Moldova (65.3), Albania (65.92), Ecuador (66.23), Peru (66.61), Brunei (66.93), Philippines (67.25), Argentina (67.28), Lebanon (67.29), Mexico (67.94), Malaysia (67.97), Turkey (68.21), Belarus (68.24), Bulgaria (68.61), Oman (69.12), Barbados (70.15), Serbia (70.29), Kuwait (70.4), Saudi Arabia (70.41), Montenegro (70.45), Thailand (70.66), Bahrain (71.07), Russian Federation (71.28), Greece (72.36), Romania (72.43), Chile (72.81), and Slovak Republic (72.88).

The highest Internet connectivity had Uruguay (73.34), The Bahamas (73.61), Latvia (73.68), China (73.98), Croatia (74.12), United Arab Emirates (74.27), Qatar (74.35), Hungary (74.71), Italy (74.86), Lithuania (75.26), Cyprus (75.99), Czech Republic (76.29), Malta (76.33), Poland (76.69), Israel (77.08), Estonia (77.71), Slovenia (77.76), Spain (79.7), France (79.87), Japan (80.04), Hong Kong

[33] https://www.gsma.com/mobilefordevelopment/resources/state-of-mobile-internet-connectivity-2018/.

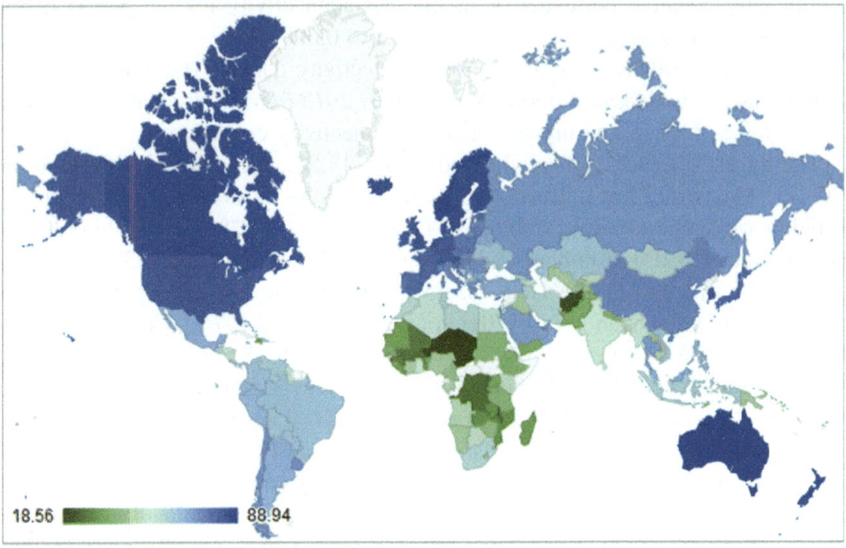

Graph 6.3 World Internet connectivity from highest (dark blue) to lowest (dark green)

(80.73), Luxembourg (81.42), Belgium (81.59), Germany (81.78), United States (81.81), Austria (82.41), Ireland (83.05), South Korea (83.37), Switzerland (83.7), Netherlands (84.16), United Kingdom (84.18), Finland (84.19), Canada (84.33), Sweden (84.33), Denmark (84.45), Norway (86.43), Singapore (86.55), Iceland (86.58), New Zealand (87.85), and Australia (88.94).

Graph 6.3 sums up the categorization into highest Internet connectivity countries in dark blue, high Internet connectivity countries in blue, low Internet connectivity countries in green, and lowest Internet connectivity in dark green based on the *State of the Mobile Internet Connectivity 2018 Index*.[34]

Graph 6.4 outlines the relation of Internet connectivity based on the *State of the Mobile Internet Connectivity 2018 Index*[35] and economic growth over time. From 2000 on, Internet-connected areas appear to show slower growth than countries with the lowest Internet connectivity.

As a cross-validation regarding the relation of GDP and AI entrance in economic markets over time, GDP growth data was derived for the years from 1961 until 2017 from the World Bank database on world economic output.[36] AI penetration of markets was measured by the proxy of the *State of the Mobile Internet*

[34] https://www.gsma.com/mobilefordevelopment/resources/state-of-mobile-internet-connectivity-2018/.

[35] https://www.gsma.com/mobilefordevelopment/resources/state-of-mobile-internet-connectivity-2018/.

[36] https://data.worldbank.org/indicator/NY.GDP.MKTP.KD.ZG.

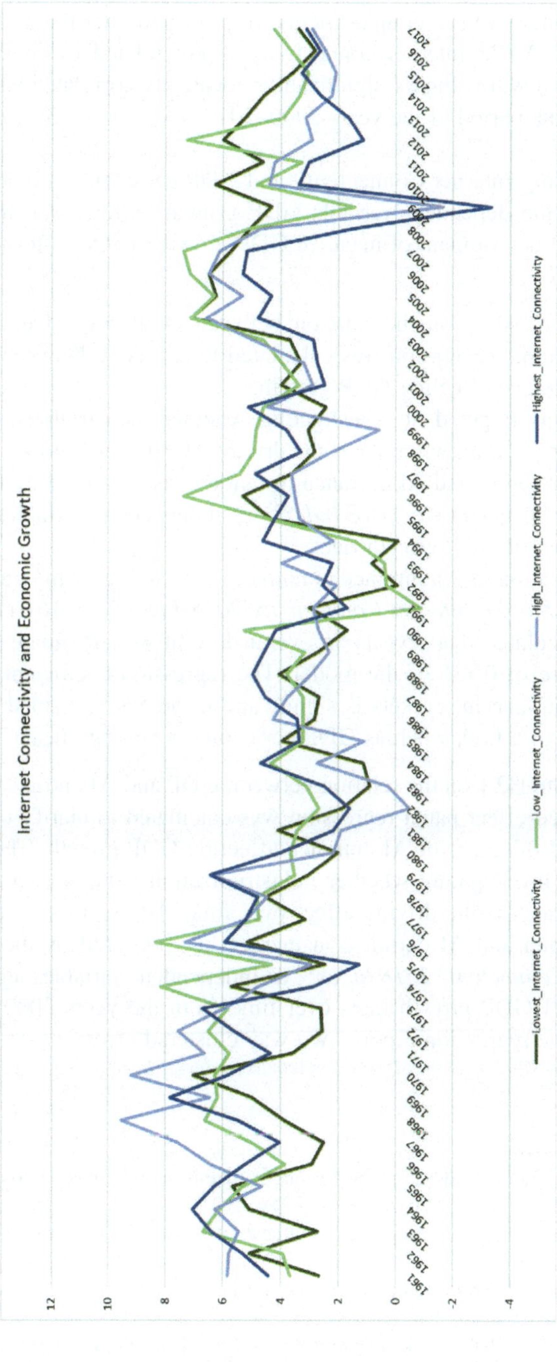

Graph 6.4 Relation of Internet connectivity based on the *State of the Mobile Internet Connectivity 2018 Index* (https://www.gsma.com/mobilefordevelopment/resources/state-of-mobile-internet-connectivity-2018/) and economic growth over time

Connectivity 2018 Index[37] of 160 countries. The world country's GDP growth over time was calculated for time compartments prior to the Internet revolution and after the Internet revolution as well as compartments for prior and after the 2008 World Financial Recession. While Internet connectivity is not related to GDP growth prior to 2000, GDP growth is highly significantly negatively correlated with Internet connectivity for the period of the years 2000–2017 ($r_{\text{Pearson}} = -375$, $n = 160$, $p < 0.000$).

A regression plotting Internet Connectivity and GDP per capita as independent variables to explain the dependent variable GDP growth outlines that the effect for AI is a significant determinant of negative GDP growth prospects for the years from 2000 until 2017.

Method: To investigate H1.$_1$ on the relation between GDP and AI entrance in markets, a cross-sectional regression was calculated to clarify in-between country differences of AI entrance affecting GDP growth.

The regression was targeted at investigating whether industrialization being associated with lower GDP growth rates is the driving effect. The regression therefore plotted GDP per capita and AI entrance in markets as measured by the *State of the Mobile Internet Connectivity 2018 Index*[38] as independent variables against the dependent variable of GDP growth rates.

To describe the relation of AI entrance in markets and GDP growth operationalized by *State of the Mobile Internet Connectivity 2018 Index*[39] of 161 countries, a regression was calculated that reveals an overall low fit with R-square of 0.147 and adjusted R-square of 0.136 of the model. The regression coefficient B value of −0.054 for AI entrance in markets is significant at the 5% one-sided t-testing level, with a p-value of 0.000, whereas GDP per capita is not significant.

Method: To investigate H2.$_1$ on the relation between GDP and AI entrance in markets over time, a fixed-effect panel regression was calculated to clarify over time in-between country differences of AI entrance affecting GDP growth. The regression was targeted at investigating whether industrialization being associated with lower GDP growth rates is the driving effect over time. The regression therefore plotted GDP per capita and AI entrance in markets as measured by the *State of the Mobile Internet Connectivity 2018 Index*[40] as independent variables against the dependent variable of GDP growth rates over time from the years 2000 to 2017 for 214 countries. The regression reports two-way clustered errors.

---

[37] https://www.gsma.com/mobilefordevelopment/resources/state-of-mobile-internet-connectivity-2018/.

[38] https://www.gsma.com/mobilefordevelopment/resources/state-of-mobile-internet-connectivity-2018/.

[39] https://www.gsma.com/mobilefordevelopment/resources/state-of-mobile-internet-connectivity-2018/.

[40] https://www.gsma.com/mobilefordevelopment/resources/state-of-mobile-internet-connectivity-2018/.

To describe the relation of AI entrance in markets and GDP growth operationalized by *State of the Mobile Internet Connectivity 2018 Index*[41] of 161 countries, a regression was calculated that reveals an overall low fit with R-square of 0.014. The regression coefficient B value of −0.059 for AI entrance in markets is significant at the 5% one-sided t-testing level, with a p-value of 0.017, whereas model 2 reveals that GDP per capita is not significant. Testing for an assumed time lag for Internet adoption of 1 year, the regression coefficient B value of −0.051 for AI entrance in markets is significant at the 5% one-sided t-testing level, with a p-value of 0.019, whereas as model 3 reveals GDP per capita is not significant as visible. Testing for an assumed time lag for Internet adoption of 1 year, the regression coefficient B value of −0.051 for AI entrance in markets is significant at the 5% one-sided t-testing level, with a p-value of 0.019, whereas model 3 reveals that GDP per capita is not significant as visible. AI entrance is negatively associated with GDP growth over time from 2000 on until 2017, whereas GDP per capita is not a relevant indicator of GDP growth.

In summary, over all the results, slowbalisation appears to be connected to human migration and differing from the currently ongoing AI revolution. Trade and transfer in data and the knowledge economy are still globalizing, while conventional globalization trends of moving goods and finance have been slowed. A current market trend toward novel technologies, such as big data capital gain, reshoring or AI taking over former human capital labor tasks and the currently-evolving 5G revolution potentially heightening this trend, is detected, which appears to be connected to GDP components of the service sector [46]. In both measurements of global connectivity as a proxy for AI entering markets from around the turn of the millennium on, higher Internet connectivity is associated with lower economic growth. This striking result demands revisiting growth theories. AI hubs are speculated to have growth—e.g., such as gains from the sharing economy, cryptocurrencies, and big data that conventional growth theory may not include. AI entrance is negatively associated with GDP growth, whereas GDP per capita is not a relevant indicator of GDP growth as was found in a cross-sectional regression and a fixed-effect country panel regression. The panel regression plotting GDP per capita and Internet connectivity from the year 2000 to explain economic growth therefore consolidates the overall finding that Internet connectivity is associated with economic growth decline over time and over 161 countries of the world, whereas GDP per capita has no significant relation with GDP growth. We may therefore advocate for revising conventional orthodox and heterodox growth theory for integrating AI-led growth with particular attention to big data gains and the 5G remote work revolution.

[41] https://www.gsma.com/mobilefordevelopment/resources/state-of-mobile-internet-connectivity-2018/.

6.2.1 Discussion

In today's economy, robots and algorithms are taking over human decision-making tasks and entering the workforce. Most recently, big data has evolved to become a source of major assets and governments around the world are endeavoring to tax wealth creation from information transfer. This trend currently challenges conventional economic theory to capture growth based on purely capital and labor components. Algorithms, machine learning, and big data gains but also the shared economy do not seem to be represented accurately in conventional growth theory components of capital and labor [5].

It is therefore proposed that contemporary growth theory should be revised as for integrating growth related to AI. First, it should be theoretically clarified, measured, and backtested on data whether AI enhances or lowers capital and/or labor components of standard growth theories. Second, as the data suggests, growth theory should consider labor to be either flexible, as would potentially be AI components, or more inflexible, as would be traditional human labor force. Third, micro-macro and endogenous and exogenous growth theories should integrate a novel component for AI as comprised of machine learning, big data, and robotics. The new growth theory proposed is

$$Yn(t) = (A(t)K(t))^{\alpha}(A(t)L(t))^{\beta}(A(t)I(t))^{1-\alpha-\beta} \tag{6.1}$$

whereby $Yn(t)$ denotes total new production function, $A(t)$ refers to capital and labor-augmenting technologies or AI knowledge, $K(t)$ is the capital, and $L(t)$ is the labor. $I(t)$ represents information, which Internet connectivity has made more accessible. Information share and big data storage as well as computation power are the most novel features of AI. Access to information but also reaping benefits from information sharing through synergizing information and deriving inferences in relation to big data is an innovative value generation in the artificial age differing from conventional capital or labor. Having already a big data collection enhances the productivity of $I(t)$ due to network effects and information being a non-rivalrous good, with a marginal utility gain that is exponential. Network effects from information and connectivity increase per additional user. Information is non-rivalrous as the consumption of one piece of information does not decrease or deplete the opportunity for another person to consume the information. The more information one holds, the better—hence, the marginal utility of information rises exponentially with information gain. In all these features—network effect gains, non-rivalrous information consumption opportunities, and exponential marginal utility gains of knowledge—information is completely different from classical notions of capital and labor. Where capital and labor are exclusive, the knowledge economy and big data-driven growth are non-exclusive [31]. A piece of information shared or written online does not take anything away or decrease utility; it actually increases people's utility non-depletable [154, 156]. Therefore, it is proposed to measure AI as completely novel component to be considered in standard growth theory. Economically, the current AI revolution is thus believed to differ from conventional technology shocks by the knowledge economy obeying different laws of economic exchange [79].

Addressing the found deficiency of an integration of AI into standard growth calculus leads to the creation of an index AI_GDP per country c based on Eq. 6.2, comprised of the GDP per capita and AI Internet connectivity percentage of a country.

$$\text{AI\_GDP}(c) = \text{GDP}_{\text{per capita}}(c) * IA(c) \tag{6.2}$$

whereby AI_GDP(c) denotes the AI-GDP index per country c calculated by GDP per capita of a country c as retrieved from a World Bank database[42] multiplied by $IA(c)$, which represents country c inhabitants' Internet usage in percent of the population as retrieved from a World Bank database.[43] Graph 6.5 tables the AI_GDP countries' indices ranked from the highest to the lowest.

Graph 6.6 AI_GDP country's index around the world. The higher the index, the darker the country is colored.

As visible in Graphs 6.7, 6.8, 6.9, 6.10, 6.11, and 6.12, continent-specific AI-GDP indices reveal Africa being relatively low on AI-GDP—see Graph 6.7. Asia and the Gulf region being in the middle ranges with Qatar and United Arab Emirates and Japan and South Korea leading as outlined in Graph 6.8. Graph 6.9 reveals in Europe Luxembourg, Switzerland, Norway, Iceland, Ireland, Sweden, and Finland as top AI-GDP countries. North America (Graph 6.10) has a higher AI-GDP index than South America (Graph 6.11), where Chile, Argentina, and Uruguay appear to lead. In Oceania, Australia has the highest AI-GDP index, followed by New Zealand as visible in Graph 6.12.

As a predicted trend, the co-existence of AI with the human species is believed to change the fundamental concepts of economic growth. Already now, we see a market disruption happening. Traditionally, globalizing areas of growth seem to stagnate, while AI-driven industries are continuously globalizing.

When considering growth theories, we may first answer the question of where AI-led growth will be driven from. AI appears as exogenous technology shock that may increase labor productivity. With this going along is a transition of the economy and legal understanding of AI. What is different in regard to AI from conventional traditional technology shocks is the missing legal framework and economic clear distinction into capital or labor. The discussion therefore covers a legal and economic analysis of what AI may represent to then propose to integrate AI as an additional growth component in growth theories.

AI has already produced novel legal creations and will do so even more in the near future, through its developing autonomy. A new legal category for AI is currently created that may instigate a new labor component in growth equations. Robots are currently partially considered as quasi-human beings in common law territories as for forming an intellectual autonomy as singular legal entities [81].

[42] https://data.worldbank.org/indicator/ny.gdp.pcap.cd.
[43] https://data.worldbank.org/indicator/it.net.user.zs.

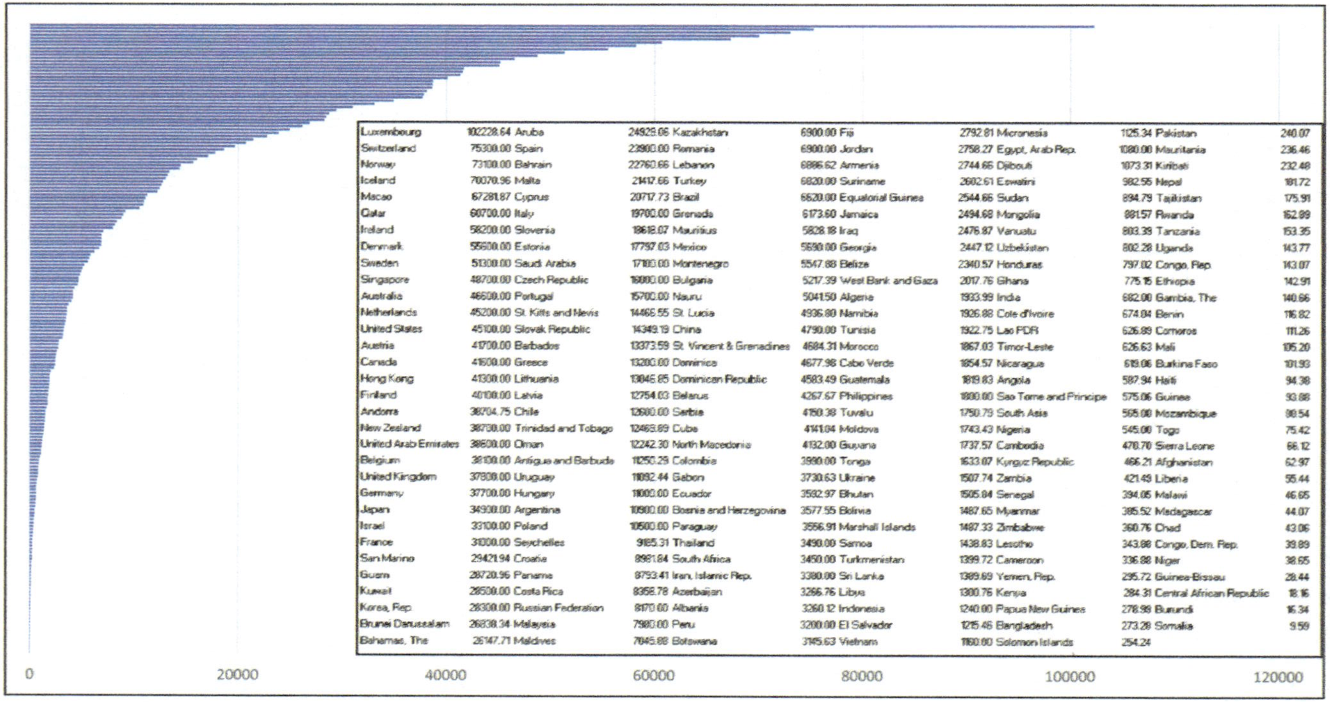

Luxembourg	102228.64	Aruba	24929.06	Kazakhstan	6900.00	Fiji	2792.81	Micronesia	1125.34	Pakistan	240.07
Switzerland	75300.00	Spain	23900.00	Romania	6900.00	Jordan	2758.27	Egypt, Arab Rep.	1080.00	Mauritania	236.46
Norway	73100.00	Bahrain	22760.66	Lebanon	6886.62	Armenia	2744.66	Djibouti	1073.31	Kiribati	232.48
Iceland	70070.96	Malta	21417.66	Turkey	6820.00	Suriname	2602.61	Eswatini	982.55	Nepal	181.72
Macao	67281.87	Cyprus	20717.73	Brazil	6620.00	Equatorial Guinea	2544.66	Sudan	894.79	Tajikistan	175.91
Qatar	60700.00	Italy	19700.00	Grenada	6173.60	Jamaica	2494.68	Mongolia	881.57	Rwanda	162.89
Ireland	58200.00	Slovenia	18618.07	Mauritius	5828.18	Iraq	2476.87	Vanuatu	803.39	Tanzania	153.35
Denmark	55600.00	Estonia	17797.03	Mexico	5690.00	Georgia	2447.12	Uzbekistan	802.28	Uganda	143.77
Sweden	51300.00	Saudi Arabia	17100.00	Montenegro	5547.88	Belize	2340.57	Honduras	797.02	Congo, Rep.	143.07
Singapore	48700.00	Czech Republic	16000.00	Bulgaria	5217.39	West Bank and Gaza	2017.76	Ghana	775.15	Ethiopia	142.91
Australia	46600.00	Portugal	15700.00	Nauru	5041.50	Algeria	1933.99	India	682.00	Gambia, The	140.66
Netherlands	45200.00	St. Kitts and Nevis	14466.55	St. Lucia	4936.80	Namibia	1926.88	Cote d'Ivoire	674.04	Benin	116.82
United States	45100.00	Slovak Republic	14349.19	China	4790.00	Tunisia	1922.75	Lao PDR	626.89	Comoros	111.26
Austria	41700.00	Barbados	13373.59	St. Vincent & Grenadines	4684.31	Morocco	1867.03	Timor-Leste	626.63	Mali	105.20
Canada	41600.00	Greece	13200.00	Dominica	4677.98	Cabo Verde	1854.57	Nicaragua	619.06	Burkina Faso	101.93
Hong Kong	41300.00	Lithuania	13046.85	Dominican Republic	4583.49	Guatemala	1819.83	Angola	587.94	Haiti	94.38
Finland	40100.00	Latvia	12754.03	Belarus	4267.67	Philippines	1800.00	Sao Tome and Principe	575.06	Guinea	93.88
Andorra	38704.75	Chile	12600.00	Serbia	4160.38	Tuvalu	1750.79	South Asia	565.00	Mozambique	88.54
New Zealand	38700.00	Trinidad and Tobago	12469.69	Cuba	4141.04	Moldova	1743.43	Nigeria	545.00	Togo	75.42
United Arab Emirates	38600.00	Oman	12242.30	North Macedonia	4132.00	Guyana	1737.57	Cambodia	470.70	Sierra Leone	66.12
Belgium	38100.00	Antigua and Barbuda	11250.29	Colombia	3980.00	Tonga	1633.07	Kyrgyz Republic	466.21	Afghanistan	62.97
United Kingdom	37800.00	Uruguay	11092.44	Gabon	3730.63	Ukraine	1507.74	Zambia	421.49	Liberia	55.44
Germany	37700.00	Hungary	11000.00	Ecuador	3592.97	Bhutan	1505.84	Senegal	394.05	Malawi	46.65
Japan	34900.00	Argentina	10900.00	Bosnia and Herzegovina	3577.55	Bolivia	1487.65	Myanmar	385.52	Madagascar	44.07
Israel	33100.00	Poland	10500.00	Paraguay	3556.91	Marshall Islands	1487.33	Zimbabwe	360.76	Chad	43.06
France	31000.00	Seychelles	9185.31	Thailand	3490.00	Samoa	1438.83	Lesotho	343.88	Congo, Dem. Rep.	39.89
San Marino	29421.94	Croatia	8981.84	South Africa	3450.00	Turkmenistan	1399.72	Cameroon	336.88	Niger	38.65
Guam	28720.96	Panama	8793.41	Iran, Islamic Rep.	3380.00	Sri Lanka	1389.69	Yemen, Rep.	295.72	Guinea-Bissau	28.44
Kuwait	28500.00	Costa Rica	8358.78	Azerbaijan	3266.76	Libya	1300.76	Kenya	284.31	Central African Republic	18.16
Korea, Rep.	28300.00	Russian Federation	8170.00	Albania	3260.12	Indonesia	1240.00	Papua New Guinea	278.99	Burundi	16.34
Brunei Darussalam	26838.34	Malaysia	7980.00	Peru	3200.00	El Salvador	1215.46	Bangladesh	273.28	Somalia	9.59
Bahamas, The	26147.71	Maldives	7645.88	Botswana	3145.63	Vietnam	1160.00	Solomon Islands	254.24		

Graph 6.5 AI-GDP index for 191 countries of the world

Graph 6.6 AI-GDP index for 191 countries of the world

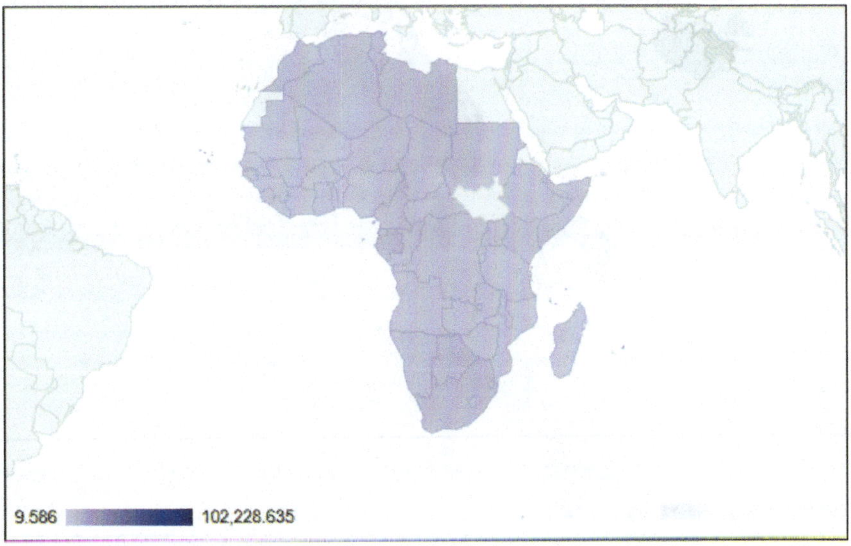

Graph 6.7 Africa AI-GDP index

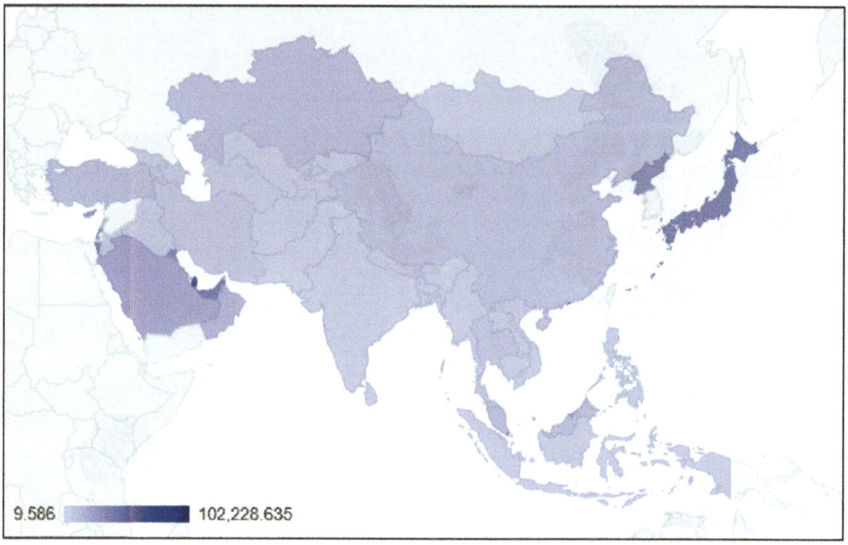

Graph 6.8 Asia AI-GDP index

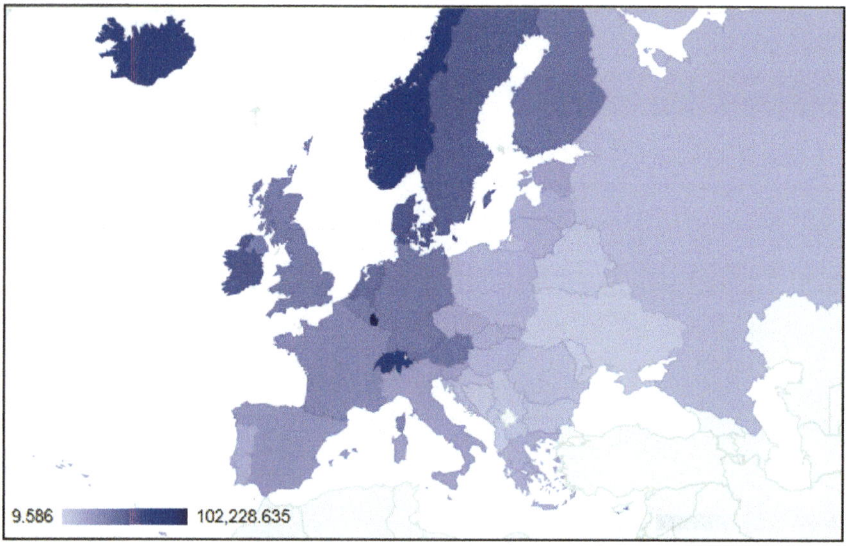

Graph 6.9 Europe AI-GDP index

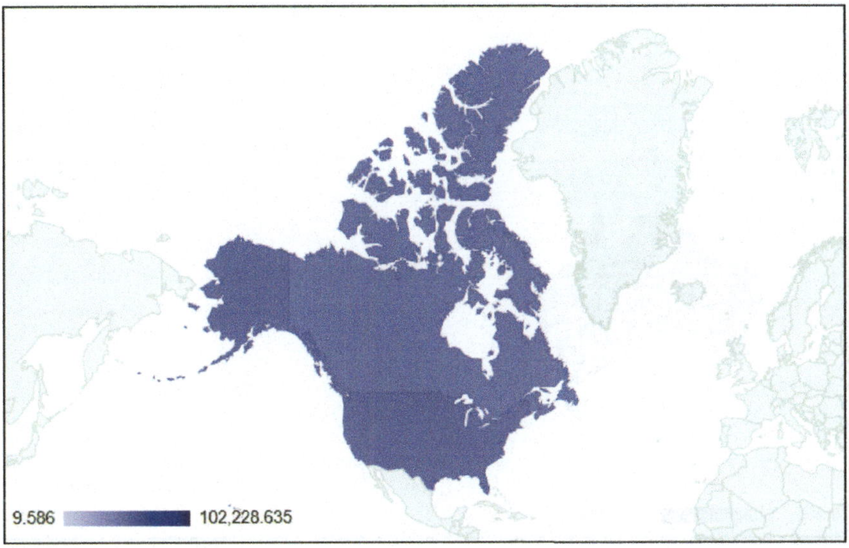

Graph 6.10 North America AI-GDP index

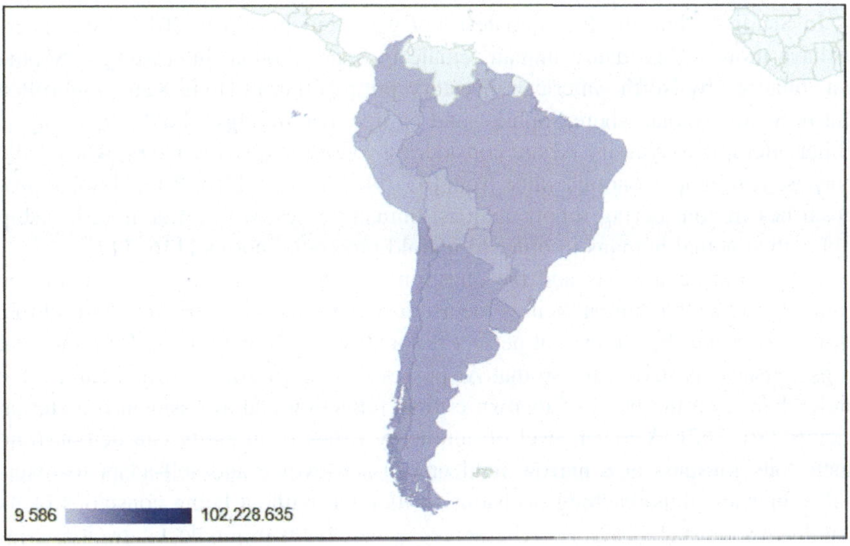

Graph 6.11 South America AI-GDP index

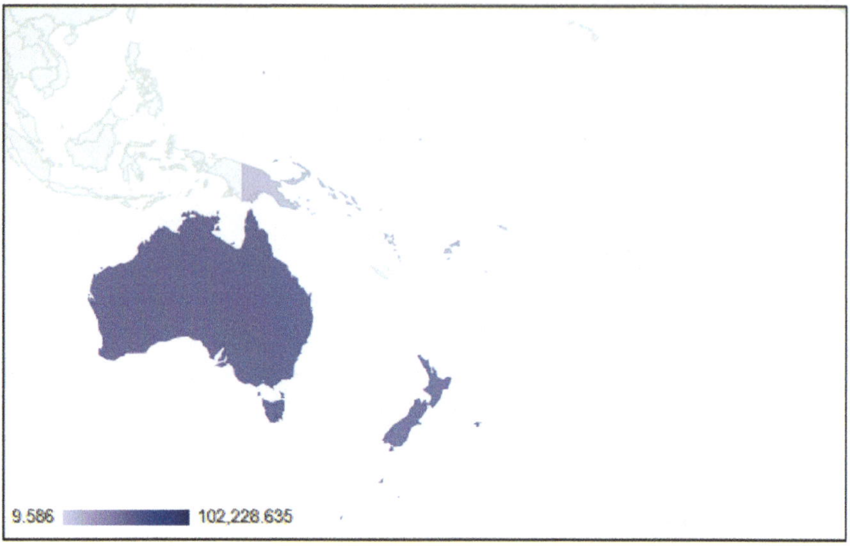

Graph 6.12 Oceania AI-GDP index

In Saudi Arabia, the first female robot got a citizenship in 2017 and appears to have more rights than a human female in Saudi Arabia. Interestingly, Sophia was financed by North American investors, put together in Hong Kong, and rolled out in Saudi Arabia, where Sophia—and by now her siblings—hold citizenship.[44] Other attempts to classify AI are considering robots as quasi-workers, whose sole purpose is to reap economic value from [35, 59, 116, 119, 120]. An additional proposal lies in considering robots as quasi-human beings, with different citizenship rights than actual humans in order to uphold pro-social norms [116, 117].

Behavioral economists add the question of whether AI and robots should be created to resemble human beings' decision-making with fast thinking and fallible choices or rather be targeted at perfect rationality and slow thinking [68]. General consciousness is strived for so that AI possesses consciousness, which can evolve and enhance on the basis of its own critical reflection and assessment of external factors [86, 162]. A lower level of autonomy exists if an entity can demonstrate such consciousness at a narrow field or can self-evolve and self-adapt to external influences, thus reaching decisions of its own, without being conscious of its intelligence as such [162]. As AI emerges as new types of intellect capacities coupled with human-like emotional features, they are attributed to a legal personhood in order to ensure to be comprehended correctly and to avoid unfair treatment, toward humans as well [162]. Respectful treatment of AI is meant to protect and uphold dignity of all people and AI [121]. Upholding certain ethics in regard to

[44] https://en.wikipedia.org/wiki/Sophia_(robot).

AI appears favorable to breed social norms but certain privileges should only be granted to human workers [69, 78, 94].

With citizenship and quasi-humanness being attributed to AI, the power relation between human and AI will need to be defined [152]. Should AI be granted full citizenship rights, the problem of overpopulation occurs, since there is the possibility of infinite life of AI. With the rise of AI persons, their eternal life poses ethical challenges in light of overpopulation and evolutionary perfection, which could crowd out human fallibility if determining merit-based eternal life. In a human-led evolution, AI will have to be switched off for various reasons, such as malfunction but also merit-based efficiency calculus. If now AI is considered as quasi-humane and granted citizenship rights, switching off AI becomes legally problematic. A human-led evolution may lead to having to decide what AI developments to favor and pursue and clear guidelines when to terminate a malfunctioning or defect AI. In this feature, AI will be different from labor for having the potential to live eternally and being more malleable to be changed and switched. AI will be flexible and interchangeable in the international arena. Again, a putty labor definition of AI components of labor is recommended that captures the difference of AI to conventional labor in this regard.

When considering the enormous physical and longevity advantages AI hold over human, a natural dominance of AI over humankind is implied. In order to ensure that human lead AI and are not subordinated, a society should be established, in which robots gain quasi-human rights but may not have the same powers and rights as human beings [166].

AI entering the workforce and holding enormous physical and longevity advantages over human but no felt emotions implies economic gains to be reaped. Standard economic growth models hold that capital and labor are essential for an economy to flourish. While capital is usually considered as fungible, exchangeable and eternal, labor is more individual, human, and inflexible. AI entering the workforce and blending in as a substitute to human capital will change the nature of labor, potentially dividing labor into a putty, flexible, eternal, and exchangeable AI part and a clay labor of inflexible human capital. In order to ensure that human can legally benefit from the economic output and growth generated by AI, a society should be established, in which robots gain quasi-human rights but may not have the same material needs and rights as human beings.

As for the international character of AI and algorithms, their fungibility, and fluid capital character, broad legal foundations of AI and the overarching regulatory framework on how to classify reaping benefits from AI should be codified in customary international law held in common among all people.

In order to protect humankind against dangers of robots and AI, negative developments should be published, stopped according to the right to destroy, and those aiding to inform about deviant developments rewarded.

AI or robots should be integrated into a blockchain as a trace on unwanted AI and robot behavior but also as a disciplinary function against other AI uprising and rebellious tendencies [112].

If AI gets legally and economically subordinated to human, ethical questions arise. According to Kant's categorical imperative, which states one should only engage in actions, one wants to be done to oneself, AI should be protected against harm and misuse or abuse. The concern here is less so the emotional and psychological state of AI, which arguably may not exist given missing self-cognition and emotions in AI, but more to set a signal and not to allow triggering sadist and negative compulsion in human that could be taken out on other humans as well, if human become conditioned and learn mistreating AI on a daily basis.

Regarding limited space on earth and sustainability concerns, longevity and eternal life of AI appears problematic. Humankind may face tough decisions whether or not to have AI proceed and what kind of developments to flourish and what to extinct [146]. In what cases should we consider to switch off AI? In 1950, Isaac Asimov introduced the idea robot to (1) not injure a human being or, through inaction, allow a human being to come to harm. (2) A robot obeying the orders given it by human beings except where such orders conflict with the first law. (3) A robot must protect its own existence as long as such protection does not conflict with the first or second law. In the cases of overpopulation and harm emerging from AI, algorithms and robots can be considered to be switched off. But when and how to stop AI?

An economic killing market mechanism may be natural market selection via price mechanisms and the falling rate of profit. Regarding prices, natural supply and demand mechanisms will always favor lucrative innovations with a higher price and the following supply of goods lead to a price drop. The falling rate of profit is one of the major underlying features of business cycles, long-term booms, and downturns [23, 24]. Capitalism is thereby described as competitive battle for innovation and reaping benefit from first-market introductions. Once followers enter the market, profit declines, leading eventually to market actors seeking novel ways to innovate in order to regain a competitive market advantage and higher rates of profit. Thereby, industries and innovations fade and die off. Such a natural market evolution is also likely to occur with AI innovations, which will determine which AI traits will remain and which ones will fade off [112, 113]. Apart from soft market mechanisms that may lead to AI evolution, what are the cases when AI should be shut down or switched off or—in the case if AI personhood—be killed?

The main and leading concern about any new and emerging technology is to be safe and error-free [88]. Therefore, sufficient and numerous tests on health and safety must be performed by developers and/or well-known independent sources before rolling out any technology onto the marketplace and society [88]. In robotics, the safety issue mainly centers around software and/or hardware designs [88]. Even a tiny software flaw or a manufacturing defect in an intelligent machine, like a smart car or a social robot, could lead to fatal results [88]. When these deviations occur and especially when they are harmful to the human community but also to other AI species, the faulty AI should be terminated. With regard to the risk of robotic malfunctions and errors, product legal responsibility laws are mostly untested in robotics [88]. A usual way to minimize the risk of damage from social robots is to program them to obey predefined regulations or follow a code-of-ethics [88]. Ethical codes for robotics are currently needed and should become formed as a natural behavioral law to then be defined and codified as law. Laws but also an ethical understanding to terminate AI, algorithms, and robots in case of impairment and harm are needed [121].

As social robots become more intelligent and autonomous and exhibit enough of the features that typically define an individual person, it may be conceivable to assign them responsibility and use them in social, educational, and therapeutic settings [88]. In the currently ongoing research on the integration of computers and robotics with biological corpse, it is found that a cognizant human brain (and its physical body) apparently has *human rights*; hence, replacing parts of the brain with artificial ones, while not harming its function, preserves those rights [88, 168]. Also, consider a handicapped person featuring an electronic robot arm that commits a crime [147]. It becomes obvious that half-robot-human beings should be considered as human and robots as quasi-human beings. Meghdari and Alemi [88] speculate that at some point in the future, we may face a situation in which more than half of the brain or body is artificial, making the organism more robotic than human, which consolidates the need of special *robot rights* and attributing (quasi)-human rights onto robots. When considering robots as quasi-human beings, their termination appears legally questionable and ethically challenging, requiring revisiting laws as legitimation to kill a likewise species as well as ethical consensus on the virtue of killing [112].

The legal argumentation may draw on justifiable homicide as outlined in criminal law cases—such as prevention of greater harm to innocents during an imminent threat to life or well-being in self-defense. According to the United Nations Universal Declaration of Human Rights, Article 3 states that everyone has the right to life, liberty, and security of person. Most nations allow for some degree of leniency for self-defense, which reduces charges. Apart from self-defense, suicide may also serve as legally justified argument for switching off AI, if artificial life is programmed to terminate itself when being harmful in such way that AI causes injury to a human being or, through inaction, allow a human being to come to harm [83].

The virtue of killing could be grounded on Viktor Mayer-Schönbergers "right to be forgotten," which ensures data privacy through automated deletion of contents after a certain period and grants individuals rights to have their data been destroyed [111–113, 121]. In this line, we may argue a "right to destroy" and program AI to stop itself should it incur hurt, damages, and losses to humankind. However, the implementation of this right is still in infancy and hindered by questions of what court is responsible for an as such claim. As a legal subsumption, we may speculate that individuals may be granted a "right to terminate" and can order for robots to be switched off if causing harm to them. As the "right to be forgotten" law can be overruled by concern for public safety, this may also apply to the right to terminate. Thereby, it deserves mentioning that safety and also expected safety standards differ around the world [112]. All these developments are prospected to lead to an AI evolution, in which humans are meant to select the process of what

AI should survive or be killed. Key decision-maker thereby diverts favorable traits and developments from unfavorable [116, 117, 119, 120, 125]. But who should determine what should survive, human or AI? A question that can be answered by sorting out the legal power relation between AI and human.

Finally, we may address the question, what is it that makes human humane and differing from AI? In the age of artificial intelligence and automated control, humanness is key to future success. Future research may draw from behavioral human decision-making insights and evolutionary economics in order to outline what makes human humane and how human decision-making is unique to set us apart from artificial intelligence rationality. Humanness as found in heuristics, decision-making errors but also procreation and creativity are believed to become more valuable in a future of AI entering the workforce and our daily lives. Drawing from behavioral human decision-making insights and evolutionary economics can help to outline what makes human humane and how human decision-making is unique to set us apart from AI rationality; AI is argued to value humanness and improve the value of human-imbued unique features [118]. All these humane features of labor should be considered as clay labor, inflexible but valuable and clearly set apart from AI.

In its entirety, the presented work-in-progress futuristic outlook promises to hold novel insights for future success factors of economic growth calculus but also human resource management grounded on efficiency and ethics. Having parts of the world being AI-driven and others being human capital grounded in the future is prospected to increase the international development divide in the years to come. While in the AI hubs human will be incentivized to become more creative and humane while AI performs all rational tasks to a maximum productivity; other parts of the world could naturally fall back as for being stuck in spending human capital time on machine-outsourceable tasks and not honing humane skills, which are not replicable by machines.

Future research endeavors may therefore address inequality drawing on the future vision that central rational and efficient AI hubs will outperform underdeveloped remote areas of the world even more in the digital age. Slowbalisation is projected to draw back outsourcing efforts and divide AI hubs from areas that are less connected. Following research should be concerned with the unprecedentedly high divide between skilled and unskilled labor and the diversion between AI hubs and non-AI territories. In the last four decades, the price of skilled labor has soared dramatically relative to that of unskilled labor despite a major uprise in the relative supply of skills. The notion of skill bias in growth theories has introduced the theoretical possibility that technological progress benefits only a subgroup of workers, placing technical change at the center of the income distribution debate [55]. Organizational changes have led to AI technologies reducing costs of communication, monitoring, and supervision within the firm, which trigger a shift toward a new organizational design. The change toward AI induces an organizational shift toward skill-biased meritocracy. Endogenous technical progress leads to economic growth, but also generates wage inequality between low- and high-skilled workers [38, 95, 101]. Faster technical change increases the return to ability and increases wage inequality between, and also within, groups of high-skilled and unskilled workers [49, 50].

Future studies should integrate some of the contemporary inequality measurements such as the Palma ratio, financial development, and wealth transfers in contemporary growth theories and measurement [64, 89, 102]. Wage inequality is only one way to assess inequality, but in order to get a richer picture of inequality derived from AI, future research may also consider inequality in wealth, health, status, and within-group inequalities [144].

Understanding the links between growth and inequality should also be placed in the different contexts of political, social, and historical environments in order to derive inference about a successful introduction of AI into today's workforce and society. Finally, more research is recommended to model and maximize the novel production function including AI and information share—especially in light of G5 and the Internet of things leading to a further connection and benefits from technology. All these novel developments may lead to a potential polarization between more efficient AI hubs and low-skilled low labor cost areas that may be shunned from economic growth due to a predicted reshoring trend coupled with AI economic dominance and unprecedented technology gains [1, 84, 85, 143, 165].

Overall, the presented work-in-progress captures AI's entrance into the workforce and our daily lives. The currently ongoing market transition of AI encroaching conventional markets will likely lead to a reordering of the current global economic and political order. The results on slowbalisation mark the very first attempt to describe slowbalisation in light of the currently ringing in AI market disruption. The findings on the relation of AI and GDP appear as first trace of AI shaping economies as if guided by an artificial drive.

Depicting growth during this unprecedented time of economic change and regulatory reform of shaping a novel technology revolution holds invaluable historic opportunities for outlining technology-driven market changes' influence on the stability of economies and society. As never before in history, automatization may enrich the world economy in very many novel ways regardless of national borders—but only if also be safeguarded by ethical imperatives. The presented research aims at the current creative destruction in the wake of AI entering the world economies being ennobled by a social face and lowering potential societal downfalls [148].

The findings may also bestow global governance policy-makers with ideas on how to better snapshot AI's potential in the digital age and market actors with future-oriented foresight how to benefit from this new technology [11, 71]. Market and societal policy recommendations may aid global governance experts to strengthen society through AI but also overcome unknown emergent risks within globalized markets in the wake of the AI revolution. At the same time of acknowledging the potential of AI, ethical considerations appear necessary as we have to become aware of the risk imbued in the artificial age, such as legal regulatory gaps and crowding out humanness or reverting the past accomplishments of outsourcing helping nations to develop out of poverty.

Conventional economic policies may therefore be coupled with a holistic vision that encompasses socio-economic and political values. Drawing attention to potential international development drawbacks and a further disparity of society based on skills and access to refined technology will offer market actors and governance bodies key insights—not only on how to benefit from a digitalizing world but also how to administer the current market transition so the benefits get distributed equally around the world. Societies of tomorrow should therefore be built on AI ethics in order to safeguard the transition to artificiality enhancing economies and ennoble society through a mutual understanding and exchange of putty and clay labor.

6.3 Big Data Ethics

Today, enormous data storage capacities and computational power in the e-big data era have created unforeseen opportunities for big data hoarding corporations to reap hidden benefits from individual's information sharing, which occurs bit by bit in small tranches over time. This chapter presents underlying dignity and utility considerations when individual decision-makers face the privacy versus information sharing predicament. Thereby, the book unravels the legal foundations of dignity in privacy but also the behavioral economics of utility in communication and information sharing. From legal and governance perspectives, the outlined ideas may stimulate the e-privacy discourse in the age of digitalization but also serving the greater goals of democratization of information and upheld humane dignity in the realm of e-ethics in the big data era [118, 122–124].

Although communication and non-communication are day-to-day decisions of individuals, to this day, there is no stringently tested utility theory of information sharing and privacy. We lack a coherent decision science framework about when people choose to share information and when they rather want to stay silent for the sake of privacy. From the economic perspective, information sharing may impose temporal irreversible lock-ins or tipping points. The point of information sharing may be a reference point, in which one bit of more communication gives less utility than one bit of less information shared; hence, one bit of more privacy grants more utility in the sense of Kahneman and Tversky's [68] behavioral decision science finding "losses loom larger than gains." There may also be a marginal decreasing utility derived from one bit more information shared but an exponential marginal utility gain from one more unit of information received given the fact that information can be put into context and an exponentially increasing marginal utility of information. Education, for instance, is the only good with an exponential marginal utility increase, as the more information one holds, the more complex connections one can make and use.

In the past, communication was depicted to decentralize organizations [33]. Media was initially promoted to offer means of information transfer, political participation, and protection against political abuse [34, 96, 97, 106, 150]. Evidence suggests that media coverage increases voter information, which increases the responsiveness of votes to policy, which increases the effort and selection of politicians, thus producing better policies [107]. Media therefore traditionally was portrayed as helping to keep politicians accountable [107]. Media coverage was found to improve selection and incentives of politicians alongside voting responsiveness [63, 150].

Critical studies in this regard show that there are negative downsides of transparency [106]. Mass media can also erode social capital, as they potentially isolate people from real-world experiences [98, 140]. A positive relation between federal funds per capita allocation to areas where the media covered political parties in power was found [150]. Research has been done on the effect of conventional media on politics including a nomenclature of biases that impose problems—especially against minority opinions [107]. Ideological biases are found in conventional media and media effects captured on vote choice [107]. While the negative facets of information on elections and the role of social media on voting outcomes have been widely discussed recently, to this day no stringent theoretical or empirical framework for the utility of privacy and information sharing on social media exits.

In the digital age, to study the trade-off between information sharing and privacy has leveraged into unprecedented importance. Social media revolutionized human communication around the globe. As never before in the history of humankind, information about individuals can be stored and put in context over time and logically placed within society, thanks to unprecedented data conservation and computational powers. The big data era, however, also opened gates to unprecedentedly reap benefits from information sharing and big data generation [109]. The so-called nudgital society was recently introduced, shedding light onto the undescribed hidden social class division between social media users and social media providers, who can benefit from the information shared by social media users. Social media users share private information in their wish to interact with friends and communicate to public. The social media big data holder can then reap surplus value from the information shared by selling it to marketers, who can draw inferences about consumer choices. The big data can also be used for governance control purposes, for instance, border protection and tax compliance control.

Drawing from the economic foundations of utility theory, this chapter seeks to introduce the first application of utility theory to a preference-value predicament between communication and privacy in the new media era. Behavioral economics insights are advanced in shedding novel light on the conflict between the humane wish to communicate now versus combined information held by unknown big data compilers in the future. An exponential loss of privacy and hyper-hyperbolic risks in the future for the information sharer is introduced as behavioral economic decision-making fallibilities. For the overconfident information sharer, it remains largely unforeseeable what the sum of the individual information sharing tranches can lead to over time and what information its Gestalt holds for those who have big data insights over time, which can also be analyzed in relation to the general population. Governance gains a critical stance on new media use for guiding on public concerns regarding privacy and information sharing in the digital age [109]. While there is some literature on the history of media on politics [107], the wide societal implications of fake news and discounting misinformation have widely been overlooked in contemporary behavioral economics research and the management as well as externalities literature. Social sciences literature on privacy and information sharing has to be reconsidered in the age of social media.

The subchapter is structured as follows: An introduction of the theory of utility and communication and information sharing is followed by an outline of the impetus of the digital big data age on privacy. The first utility theory of information sharing and privacy will be theoretically introduced. Hyperbolic decision-making fallibility will become the basis of argumentations around hyper-hyperbolic discounting—the novel argument that information sharing in tranches may lead to an underestimation of the privacy infringements when these bits of information can be put together over time and are compared to big data in order to infer about the individual in relation to the general population.

The subjective additive utility of information shared tranche by tranche may underestimate the big data holder's advantage to reap benefits from information shared. The discussion introduces problems of the contemporary nudgital society [109], in which big data compilers can reap a surplus value from selling compiled information (*The New York Times*, November 14, 2017)[45] or manipulate market behavior or reap benefits from vulnerable population segments based on their previously shared information (*The Economist*, November 4, 2017).[46] Implications lead to open questions about ethics in the information age and recommendations for a reclaiming of the common good of shared knowledge in education about information sharing in the digital age as well as the democratization of information. Challenging contemporary behavioral insights theory aims at fostering a more informed, self-determined, and protected digital society in the wish to uphold e-ethics in the twenty-first-century big data social media era.

Implications lead to open questions of ethics in the information age and the protection of consumers with special consideration of vulnerable groups, such as patients. For example, in the course of the novel Coronavirus pandemic, many sick people with mild symptoms turned to self-help groups on the Internet outside of hospitals to search for information. In doing so they often made sensitive information about their state of health publicly available on the Internet. This information could lead to selection disadvantages for future application processes or even spouse selection preferences.

Recommendations point to a democratization of gains and insights from jointly developed knowledge available online during the collective exchange of information in the digital age. The contemporary theories of behavioral research and behavioral analysis should also promote a better informed, self-determined and

---

[45] https://www.nytimes.com/2017/11/14/business/dealbook/taxing-companies-for-using-our-personal-data.html?rref=collection%2Fsectioncollection%2Fbusiness&action=click&contentCollection=business&region=stream&module=stream_unit&version=latest&contentPlacement=8&pgtype=sectionfront.

[46] https://www.economist.com/news/leaders/21730871-facebook-google-and-twitter-were-supposed-save-politics-good-information-drove-out.

protected digital society in order to improve the e-ethics of online social media and big data information gains.

6.3.1 Utility

Economic theory is built upon the idea of utility, which captures people's preferences or values [44]. Humans are believed to strive to maximize utility on a constant basis by weighting their preferences and values on the pleasure they would receive from different options. In neoclassic economics, utility theory primarily focuses on prescriptive utility maximization giving recommendations on how individuals should behave to maximize their utility. Prescriptive utility maximization theory serves as normative guide in helping the decision-maker codify preferences. If preferences would violate rational preference choices, the theory suggests strategies so the informed decision-maker can revise their rational reference choices and judgments to eliminate preference inconsistency.

Using utility theory, preferences are constantly transformed into corresponding numerical utility data that is portrayed to maximize the individual's pursuit of happiness. Utility theory provides a powerful set to determine how to compare actual alternatives. It enables the decision-maker's optimal preferences to be transformed into a numerical utility structure guided by an optimization algorithm.

In doing so, standard microeconomic utility theory has been of aid to explain how to maximize individual outcomes in very many different domains ranging from marketing research [56, 82, 153], food industry quality control of products and corporate strategies [142, 155], and production [8, 9, 157].

6.3.2 Dignity

Dignity is the right to be valued and respected for one's own sake and to be treated ethically. Everyone has a right to respect for their dignity.[47] As an inherent, inalienable right, dignity is a core concept in fields such as morality, ethics, law, and politics. Often connected to identity and respect for integrity and other fundamental freedoms and rights, dignity is often used to uphold the ethical considerations around oppressed and vulnerable groups, who do not have insights about the consequences of their actions.

Individuals derive self-worth from dignity. While dignity itself seems to be a vague concept, it is often used as a boundary condition of what is right, just, and fair to argue for the improvement of conditions for discriminated, vulnerable, and targeted. Violations of dignity are felt as humiliation, instrumentalization or objectification, degradation, and dehumanization.

---

[47] United Nations 1998 *UNESCO Declaration on the Human Genome and Human Rights*. At Article 2.

Privacy infringements have been argued to hold concerns regarding dignity. In the age of digital media and big data, when individual decision-makers may have hyper-hyperbolic discounting fallibility regarding their share of data, dignity infringements may happen mainly unnoticed. Individuals may be endowed with reason and conscience but their decision-making capabilities may not be able to discount the worth of their information in the future and in relation to other individuals' data forming big data insights.

The freedom of expression may hold a shadow of the future. Dignity-based protection of medical patients and in biology settings may serve as dignity-based privacy beacons in the age of big data. Dignity has become the legal–ethical foundation of new reproductive and genetic technologies, medicine and genetic ethics research on humans, life and health sciences, ethics around cloning, medical integrity, bioethics, but also against war cruelty, criminal punishment, imprisonment, terrorism, weapons, abortion, sex work, and defamation. The core idea of dignity is prevalent in cultures of the world and has been extended onto animals and the environment.

6.3.3 Information Sharing and Privacy

The wish for communication is inherent in human beings as a distinct feature of humanity. Information transfer is particularly important during times of fast-paced change and extraordinary levels of uncertaintly. The COVID-19 pandemic as an unexpected external shock accounts for a major life-changing event that may shape Generation COVID in the decades to come.

Leaving a written legacy that can inform many generations to come is a humane-unique advancement of society. At the same time, however, privacy is a core human value. People choose what information to share with whom and like to protect some parts of their selves in secrecy. Protecting people's privacy is a codified virtue around the world to uphold the individual's dignity. Yet to this day, no utility theory exists to describe the conflict arising from the individual preference to communicate and the value of privacy.

6.3.4 The Humane Preference for Communication

The act of conveying intended meanings from one entity or group to another through the use of mutually understood signs and semiotic rules is the act of communication. Communication is a key feature of humans, animals, and even plants [169]. Steps inherent to all human communication are the formation of communicative motivation and reason, message composition as further internal or technical elaboration on what exactly to express, message encoding, transmission of the encoded message as a sequence of signals using a specific channel or medium, noise sources influencing the quality of signals propagating from the sender to one or more receivers, reception of signals and reassembling of the encoded message from a sequence of received signals, decoding of the reassembled encoded message, and interpretation or sense-making of the presumed original message [149].

Information sharing implying giving up privacy is at the core of communication. Communication can be verbal and nonverbal. Comprising very many different domains ranging from business, politics, interpersonal, social to mass media, communication is a humane-imbued wish and center core of every functioning society.

In society, language is used to exchange ideas and embody theories of reality. Language is the driver of social progress [99]. Linguists find discourse and information sharing inseparable from socio-economic societal advancement [48]. Language and communication modes are implicit determinations of social strata [99].

Different institutions and media sources have different varieties of language and information sharing styles. Access to information is related to social status and market power. Social visibility is a powerful and cheap incentive to make people contribute more to public goods and charities and be less likely to lie, cheat, pollute, or be insensitive and antisocial [3]. Information receipt is an implicit determinant to classify and rank people to assert institutional or personal status in society [48]. Mass communication echoes in economic cycles in the creation of booms and busts [134]. Media is also a hallmark of propaganda and political control [22, 107]. At the same time, privacy is a human virtue around the world.

6.3.5 Privacy as a Human Virtue

Privacy is the ability of an individual or group to seclude themselves, or information about themselves, and thereby share information about themselves selectively. The right to privacy grants the ability to choose which information about parts of the self can be accessed by others and to control the extent, manner, and timing of the use of those parts we choose to disclose. Privacy comprises the right to be let alone, the option to limit the access others have to one's personal information, and secrecy as the option to conceal any information about oneself [151].

The degree of privacy varies in autonomy levels throughout individualistic and collectivism cultures. While the boundaries and contents protected and what is considered as private differ widely among cultures and individuals, the common sense in the world is that some parts of the self should be protected as private.

Privacy has a valued feature of being something inherently special or sensitive to a person, which can create value and specialty if shared with only a selected person or group. The domain of privacy partially overlaps with security, confidentiality, and secrecy, which are codified and legally protected throughout the world, mainly in privacy laws but also in natural laws of virtues of integrity and dignity.

Privacy is seen as a collective core human value and fundamental human right, which is upheld in constitutions around the world[48] [66, 167].

In personal relations, privacy can be voluntarily sacrificed, normally in exchange for reciprocity and perceived benefits. Sharing private information can breed trust and bestow meaningfulness to social relations. Giving up privacy holds risks of uncertainty and losses, which are undescribed in economics and in particular the behavioral economics literature on intertemporal decision-making [51]. People tend to be more willing to voluntarily sacrifice privacy if the data gatherer is seen to be transparent as to what information is gathered and how the information will be used [100].

Privacy as a prerequisite for the development of a sense of self-identity is a core of humanness [4]. Privacy is often protected to avoid discrimination, manipulation, exploitation, embarrassment, and risks of reputational losses, for instance, in the domains of body parts, home and property, general information of private financial situations, medical records, political affiliation, religious denomination, thoughts, feelings, and identity.

Technological shocks have a history of challenging privacy standards [167]. The age of instant messaging and big data, however, has leveraged the idea of privacy to another dimension. The concept of information privacy has become more significant as more systems controlling big data appear in the digital age. With advances in big data, face recognition, automated license-plate readers, and other tracking technologies, the upholding privacy and anonymity have become increasingly expensive and the cost of upholding privacy is opaquer than ever before [3].

6.3.6 Privacy in the Digital Big Data Era

The amount of big data stored each second has reached an all-time high in the digital era. Internet privacy is the ability to determine what information one reveals or withholds about oneself over the Internet, who has access to personal information, and for what purpose one's information may be used. Privacy laws in many countries have started to adapt to changes in technology in order to cope with unprecedented constant information surveillance possibilities, big data storage opportunities, and computational power peaks. For instance, Microsoft reports that 75% of U.S. recruiters and human resource professionals use online data about candidates, often using information provided by search engines, social network sites, photo and video sharing tools, personal web appearances like websites and blogs, as well as Twitter. Search engine algorithm data maintenance has become

[48] For example, Asian-Pacific Economic Cooperation, Australia, Brazil, Canada, China, European Union, Italy, Japan, Korea, Organisation for Economic Co-operation and Development, South Africa, United Kingdom, United Nations, United States, Universal Declaration of Human Rights—to name a few.

a market advantage when thinking of Search Engine Optimization (SEO) services that help create positive content on the Internet. Google and other search engines have set themselves the task of correcting targeted misrepresentation on the Internet.

Social media tools have become large-scale factories with unpaid labor [109]. For instance, Facebook accounts for the largest social network site with nearly 1490 million members, who upload over 4.75 billion pieces of content about their lives and that of others daily. The accuracy of this information also appears questionable, with about 83.09 million accounts assumed to be fake. Aside from directly observable information, social media sites can also easily track browsing logs and patterns, search queries or secondary information giving inferences about sexual orientation, political and religious views, race, substance use, intelligence and overall personality, mental status, individual views, and preferences [72, 73].

As for the unprecedented possibilities to collect data, store big data, and aggregate information that can be compared to big data Gestalt over time and society, privacy has leveraged into one of the most fragile areas of concern in the electronic age, demanding legal protection, regulatory control, and e-ethics [45]. Today, the existing global privacy rights framework in the digital age has been criticized to be incoherent, inefficient, and need for revision. Global privacy protection shields are demanded to be established. Yet to this day, there is no economic framework on information sharing and privacy control. While—for instance—Posner [103] criticizes privacy for concealing information, which reduces market efficiency, Lessig [77] advocates for regulated online privacy. As of now, we lack a behavioral decision-making frame to explain the privacy paradox of the individual predicament between the humane-imbued preference to communicate and information share versus the value of privacy. We have no behavioral economics description of inconsistencies and moderator variables in the decision between online information sharing behavior and retroactive preference reversal preferences in the eye of privacy concerns in the digital big data era. Our understanding of strategic searchplace manipulation in SEO is growing on importance but—to this day—hardly related to behavioral environmental nudge strategies and rising online ethicality demands.

6.3.7 A Utility Theory of Privacy and Information Sharing

Building on classical utility theory, individuals are constantly evaluating competing choice options. Individuals weigh alternative options based on their expected utility derived. Indifference curves would then connect points on a graph representing different quantities of two goods, between which an individual is indifferent.

In the case of the privacy paradox of information sharing preferences and privacy values, a person would weigh whether or not to share information s or choose the information to remain private p. The respective indifference curves would outline how much of information sharing s and privacy p can be enabled to end with the same utility given the budget of overall information held by the decision-maker.

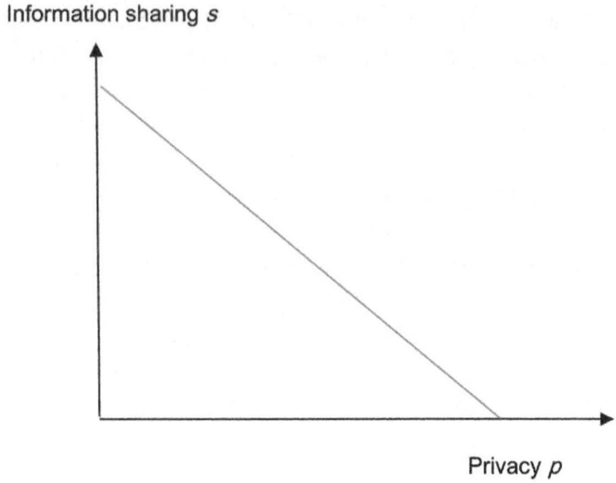

Graph 6.13 Indifference curve (blue line) for information sharing s and privacy p given the total information and communication constraint

Graph 6.13 represents the respective indifference curves for information sharing s and privacy p. That is, the individual has no preference for one combination or bundle of information sharing or privacy over a different combination of the same curve. All points on the curve hold the same utility for the individual. The indifference curve is therefore the locus of various points of different combinations of privacy and information sharing providing equal utility to her or him. Indifference curves are thereby seen to represent potentially observable behavioral patterns for individuals over information bundles. The indifference curve for information sharing s and privacy p is subject to communication and information constraints, hence, all information budgets and communication opportunities. There is only a finite amount of information, and there may be environmental conditions determining if people can exchange and share information. As exhibited in Graph 6.13, the indifference curve for information sharing s and privacy p is a straight line given the assumption that information sharing or privacy is substituted.

While, in classical economics, an individual was believed to always being able to rank any consumption bundles by order of preference [65],[49] the indifference curve for information sharing s and privacy p subject to communication and information constraints may feature a hyper-hyperbolic element or temporal dimension. The information share moment may thereby be a reference point. At the moment of the information sharing decision, it may not be foreseeable what the future implication of the information sharing is.

In general, the costs and benefits of communication are assumed as linear subtraction of positive benefits of communication b_c minus the negative consequences of communication c_c. The nature of the problem is intertemporal as information sharers cannot foresee the future implications of their information sharing divided

[49] http://www.econlib.org/library/YPDBooks/Jevons/jvnPE.html.

by variance σ [104]:

$$\frac{b_c - c_c}{\sigma} \tag{6.3}$$

However, the digital social media era has heralded a hyper-hyperbolic discounting fallibility. Individuals have lost oversight of the consequences of their individual information sharing given big data hoarding capabilities, which also allow drawing inferences about the individual in relation to others.

In the digital big data era, information share online may hold unforeseen risks of privacy merchants or social media capitalists that commercialize information reaping hidden benefits from the information provided [43, 109] (*The Economist*, November 4, 2017).[50] The subjective additive utility of information shared tranche by tranche may underestimate the big data holder's advantage to reap benefits from information shared given unprecedented data storage and big data computation power advantages of the big data era. Unprecedented computational power and storage opportunities have created the possibility to hoard information over time and put it in context with the rest of the population in order to draw inferences about the information sharer (*The New York Times*, November 14, 2017).[51]

The digital age and era of instant information sharing have therefore heralded problems of individuals who give in their basic humane need for information communication to become vulnerable over time. The big data information holder may thereby benefit from the history of information and the relation of the individual's information in comparison with the general population to an unknown degree, given missing e-literacy and transparency. Comparison to the general public may lead to an implicit underrepresentation and hence discrimination of vulnerable groups. For instance, certain groups that may not be represented online will therefore likely face an under-advocacy of their rights and needs.

While regular hyperbolic discounting captures a game-theoretical predicament of the self now versus the self later, the information offering more of a Gestalt in the eyes of the big data holder, leverages hyperbolic discounting to a game theory against uncertainty on the end of the big data holder. The hyper-hyperbolic discounting fallibility therefore may describe that at the moment of information sharing, the individual has hardly any grasp of what is implied in the giving up of privacy. The individual only focuses on the current moment trade-off between information sharing and privacy upholding, but hardly has any insights into what the compiled information over time holds for big data moguls. As for holding computational and storage advantages, the social media big data moguls can form a

[50] https://www.economist.com/news/leaders/21730871-facebook-google-and-twitter-were-supposed-save-politics-good-information-drove-out.

[51] https://www.nytimes.com/2017/11/14/business/dealbook/taxing-companies-for-using-our-personal-data.html?rref=collection%2Fsectioncollection%2Fbusiness&action=click&contentCollection=business&region=stream&module=stream_unit&version=latest&contentPlacement=8&pgtype=sectionfront.

Gestalt which is more than the sheer sum of the individual information shared, also in comparison with the general populace's data. The shared information can also be resold to companies [43] (*The New York Times*, November 14, 2017).[52] In relation to other people's information, the big data moguls can make predictions about their choices and behaviors.[53] Information can also be used for governance purposes, for instance, tax compliance and border control mechanisms [109]. Some governments have recently used big data to check the accuracy of tax reports but also to detect people's political views when crossing borders [109]. Lastly, the use of big data inferences also implies hidden persuasion means—nudging can be turned against innocent information sharers who have no long-term and computational advantage to foresee the impact of the information share (*The Economist*, November 4, 2017) [109].[54]

While behavioral economics hyperbolic discounting theory introduces the idea of time inconsistency of preferences between an individual now and the same individual in the future, hyper-hyperbolic discounting underlines that in the case of information sharing preferences this fallibility is exacerbated. In the digital context, individuals lose control over their data as big data moguls can reap surplus value from the social media consumer-workers' information sharing. Big data amalgamating social media platform companies can derive information complied over time and put it in relation to the general norm in to draw inferences about the innocent information sharer. With the modern digital era, all these features open an information sharer versus information reaper divide in the big data age [109].

From the social media big data capitalist view, the information gain of one more person sharing information is exponentially rising. Hence, the marginal utility derived from one more person providing information on social media platforms is increasing exponentially and disproportionally to the marginally declining costs arising from one more person being added to the already existing social media platform. Communication costs and benefits are assumed to not be additive and separable.

[52] https://www.nytimes.com/2017/11/14/business/dealbook/taxing-companies-for-using-our-personal-data.html?rref=collection%2Fsectioncollection%2Fbusiness&action=click&contentCollection=business&region=stream&module=stream_unit&version=latest&contentPlacement=8&pgtype=sectionfront.

[53] https://www.nytimes.com/2017/11/14/business/dealbook/taxing-companies-for-using-our-personal-data.html?rref=collection%2Fsectioncollection%2Fbusiness&action=click&contentCollection=business&region=stream&module=stream_unit&version=latest&contentPlacement=8&pgtype=sectionfront.

[54] https://www.economist.com/news/leaders/21730871-facebook-google-and-twitter-were-supposed-save-politics-good-information-drove-out.

6.4 Advances in Behavioral e-Economics and e-Ethics

Advances in the digital behavioral economics age are likely to show e-ethics trends. E-ethics in the digital century focus on inclusion in the wake of social justice pledges that serve a demand on comparative approaches to understand the most contemporary responsibility challenges of our time. With the COVID-19 pandemic having exacerbated existing inequalities and rising new gaps within society, inequality alleviation will become essential in the post-COVID-19 era in the domains of access to affordable healthcare, finance, education, digitalization and sharing the burden to protect the environment. Law and Economics developments may aid in envisioning a transition to a more inclusive society. While the legal analysis grants insights about the disparate impact of policies, the economic analysis allows to study efficiency of burden sharing over time and with consideration of externalities.

6.4.1 Post-COVID Inequality

Digitalization offers unprecedented human advancement and democratization potential free from corruption. At the same time, shifting marketplaces to online virtual spaces opens gates for misinformation and disinformation being used in a competitive sense. Ethics of inclusion, Law and Economics advocacy and interdisciplinary dialogue building but also human-artificial intelligence algorithm compatibility are expected to become key advancements in behavioral economics and finance leadership of the future.

In the most recent decades, human decision-making heuristics were studied to show how nudging and winking can help citizens to make rational choices. Behavioral economics started with outlining human decision making deviations from rationality, so-called heuristics or mental shortcuts. Heuristics were perceived as failures in the North American Behavioral Economics School, while the European tradition saw human decision making as successful strategy to cope with an overly complex world. Mental heuristics and biases were studied in field and laboratory experiment but also by the help of big data and online observations to retrieve powerful nudges to benefit from life and economic markets. Over time, a broad range of nudges and winks were developed to curb the harmful consequences of human decision-making or improve human fitness to adapt to the environment. Some techniques were communicated openly, while other behavioral insights informed more subliminal change strategies.

Since a bit more than 10 years, the behavioral economics approach was applied to political context in the behavioral insights can be employed for the greater societal good. Most recently, leadership and followership directives on nudging in digitalized decision-making and the use of nudging and winking in the digital age. Behavioral insights to overcome inequality in the post-external shock changing every sphere of human being [53, 136].

The COVID-19 pandemic is likely exacerbating existing inequalities between the finance world and the real economy given the financial market performance capabilities to exchange losing for winning industries during downturns and the real economy dependence on wages and trickle-down economic effects [129, 130]. In addition, homeownership and inflation-adjusted salary schemes become a driver of inequality in our longest-lowest-ever inflation regime [130].

COVID-19 also made apparent vast differences in healthcare provision around the world and the country differences in quality care [132, 135]. Within workplace settings, digitalization will become essential and drive an already-existing gap between e-skilled and e-unskilled labor [53, 129, 138]. All these rising inequalities will drive the currently-unfolding demand for ethics of inclusion as the spring feather of equality and social justice heralding in our post-pandemic Renaissance [129, 132].

In the economics and finance sectors, inequality alleviation could bridge the performance gap between the finance world and the real economy with particular attention to inflation, low interest rate regimes and metropolitan versus rural areas. Disparate impact analyses will pave the way for a more diversified scheme of financial rescue and recovery aid. In the healthcare domain, the equal access to medical care pledge in innovations—such as telemedicine and artificial intelligence, robotics and big data insights—should be endeavored. Health and well-being underlying human workforce productivity can be introduced as a hidden driver of economic growth in the eye of a global pandemic risk. Digital healthcare can bring along access to affordable quality care around the world free from corruption [132]. Digitalized healthcare information exchange has nowadays also proved as a means for a democratization of health and disease remedies when considering the role of online media forums to discuss COVID long-haul issues [136].

Ethics of inclusion of the future may advance online access to healthcare and a fair and protected information exchange of vulnerable patients. Our workplace revolution in a truly digitalized economy will also drive novel inequalities and ethical dilemmas arising from digitalization. In addition, educational social transfer hubs with attention to online opportunities have leveraged into a gateway of social justice transformation in the US that also account for the most promising international development advancement of our times [32]. Environmental demands for a transition to a green economy are met in most novel economic attempts to stimulate the economy under the umbrella of sustainability and social inclusion—such as outlined in the Green New Deal and European Green Deal including a Sustainable Finance Taxonomy [127].

Future behavioral economics advancements and finance leadership will include the use of behavioral insights to improving the societal welfare under these novel frameworks. Behavioral insights for a greening of the economy will include micro-economic foundations but also build on a growing body of behavioral macro-economic findings of the unequal distribution of climate change gains and losses [128]. Law and Economics Disparate impact analysis of law and economics interdisciplinary endeavors account for the most cutting-edge novel advancements in behavioral sciences [131]. On the macrolevel, academic field of Law and Economics offers legal long-standing excellence on accounting for disparate impact, inequality and redistribution. The analysis of macro-economic aggregates benefits from a legal scholarship-led reflection of diversified and temporal views of social preferences, given patterns of differences based on age, gender, race and professional propensity risks.

The economic inequality exacerbated by COVID-19 implies that future economic policy research may take inspiration from the legal concept of disparate impact to channel the currently unprecedentedly-large rescue and recovery aid wisely to alleviate inequality [130]. Measuring a potentially disparately-heavy impact of external shocks—such as COVID-19 but also climate change or racial disparities—on gender, race and other stratifying classifications that may lead to discrimination open gates for targeted rescue and recovery aid with particular attention to empowering minorities and/or alleviating disadvantages.

Already now we see a pegging of governmental rescue and recovery aid to socially-uplifting causes to address inequality concerns and environmental causes [128, 130]. Future efforts could directly investigate if there is a heavier load of the societal burden due to COVID-19 on particular groups that hinder a speedy, full and fair recovery. Legal excellence on how to detect disparate impacts could be coupled with behavioral insights on how to alleviate biases via taxation redistribution in an uncertain world in order to rescue, uplift and empower weaker societal segments in the age of our post-COVID Renaissance.

6.4.2 Searchplace Discrimination

As for searchplace discrimination, digitalization offers unprecedented human advancement and democratization potential free from corruption [133]. For instance, in the medical sphere, telemedical assistance helps bring access to quality care and information about disease control to remote areas of the world or countries that are struck by corruption. Access to crowdsourcing information in online forums about novel diseases and alternative medical solutions for chronic debilitation is a groundbreaking phenomenon of our times [15, 136].

At the same time, shifting marketplaces to online virtual spaces opens gates for misinformation and disinformation in search engines and online forums being used in a competitive sense. The strategic manipulation of environments is a feature of behavioral economics. But while the focus is primarily on how to use nudges and winks to make the world a better place in helping humans make wiser decisions; less is written and known about how unethical behavior is used to curb and distort online environments by deleting information or clogging online searchplace users with unnecessary misinformation or compromising disinformation. Less documented and not regulated are behavioral black hat strategies that have become used competitively in searchplaces, such as Google, Yahoo or Bing.

In the most recent decade, searchplaces, such as Google, Yahoo or Bing, have gained prominence for screening and scanning candidates in workplace related contexts. While we have a most important and emerging stream of literature on algorithmic discrimination and inherent biases in searchplaces and internet forums [173]; less is known about strategic searchplace distortion. So-called Search Engine De-optimization black hat strategies are competitive and unethical distortions of search engine results that either overemphasize unfavorable search results (enabled via clickfarms) or create a misinformation or disinformation overload that derails from accurate representations of individuals online. For instance, Google search results can get capped at a low number (indicated by Google) and unfavorable, misleading and/or compromising information can get highlighted via clickfarms. Positive or content information can also be erased by flagging content online that gets immediately taken offline and hardly any resuscitation control by human reviews is installed.

While there is a possibility to craft the internet search results together and flag and report inappropriate online content, little quality control is given to this option being used as a strategy to push down or make content disappear that is appropriate in the wake of competition. For instance, competitors could use the flag and report button to make content of competitors unavailable. While the early 2000s was a period of advocacy for the 'Right to Delete' [87], the 2020s should also thematize the 'Right to not be Deleted.'

Potentially quality control over what gets flagged and who manipulates search results strategically could be enacted. Legal advancements should include clear guidelines and oversight of fraudulent use of the internet in a competitive way that manipulates genuine algorithm results, for instance via clickfarms or backlinks that curb or tilt search results in a particular way. Lastly, those, who face a searchplace discrimination disadvantage should be protected by legal regulation, technical support and rescue funds established by the industry to uphold to favor quality over unethicality in their profession. After all, sensitivity for cyberbullying via misinformation and disinformation could help elevate professions to a more ethical ground and uphold focus on excellence and merit rather than discreditation potential due to unethical conduct and lacking human algorithmic control.

6.4.2.1 Regulating searchplace discrimination

With instant information exchange on social online media playing an increasing role in politics, finance and economics, governments all over the world have set out to better understand the collective social impact of online content.[55] Social crowd control through online information display has become subject to debate of governmental security.[56]

[55] New Zealand Prime Minister Jacinda Ardern, Harvard Commencement Speech 2022, May 26, 2022, YouTube, Retrieved at https://www.youtube.com/watch?v=xOg7FJBBbJc

[56] Speech by Foreign Minister Annalena Baerbock, Seizing the Transatlantic Moment: Our common responsibility in a new world, The New School, New York City, Aug. 2, 2022, Twitter, Retrieved at https://twitter.com/i/broadcasts/1YpKkZlbLgAxj James A. Green, The Rise of Twiplomacy and the Making of Customary International Law on Social Media, Chinese Journal of International Law, 21(1), 1–53, https://doi.org/10.1093/chinesejil/jmac007

The role of social media online information exchange for law making, in particular in developing customary law in the international law context, has recently become subject to scrutiny by the International Law Commission and debate in global governance and policy contexts.[57]

In the comparative international arena, governance and governmental approaches to regulate internet activities differ substantially around the world. Even with the complex supranational legal system of the European Union, the approaches of individual countries in curbing the global phenomenon of internet security vary.

The governmental and regulatory responses to individual searchplace discrimination, however, are to this day just forming in response to industry developments. Individual searchplace discrimination via SEDO techniques and black hat strategies are yet to this day fairly undiscussed and unregulated. One reason for this may be that the changes to search results often appear only temporally and/or gradually so that the manipulation is barely noticed and hard to track down. Discourse on these tactics remains scarce, and legal and regulatory measures appear insufficient.

In general, the growth of SEO has resulted in consumer pressure to debias searchplaces. In recent years, the public opinion and stakeholder critique made – for instance – the market leader Google pay increasing attention to correcting errors and manipulative distortions.[58] With Google management having become aware of deliberate distortions of search results and increasing stakeholder concern over the credibility and misuse of online searchplaces, industry responses against strategic, competitive manipulation have turned search engine providers to punish strategic manipulation of search engine results.[59]

Technically, search engine engineers, like Google and Bing, can change the position of a website if ranked very low compared to its competitors but it requires human attention and likely compulsory regulatory action for searchplace providers to do so.[60] If being called for action and detecting misuse, Google can immediately remove spam and harmful contents from Google searches and also correct misleading snippets. Google also informs that if strategic manipulation of search

[57] James A. Green, The Rise of Twiplomacy and the Making of Customary International Law on Social Media, Chinese Journal of International Law, 21(1), 1-53, https://doi.org/10.1093/chinesejil/jmac007

[58] Official Google Webmaster Central Blog: A quick word about Googlebombs, Googlewebmastercentral.blogspot.com. January 25, 2007 (Retrieved July 9, 2010); Jacqui Cheng (January 26, 2007), Google defuses Googlebombs, News: Ars Technica (Retrieved January 27, 2007); https://en.wikipedia.org/wiki/Search_engine_results_page

[59] Mayer, Marissa (September 16, 2005). Official Google Blog: Googlebombing 'failure', Googleblog.blogspot.com (Retrieved July 9, 2010); Cohen, Noam (January 29, 2007), Google Halts 'Miserable Failure' Link to President Bush, The New York Times (Retrieved May 3, 2010); Urban Legends Reference Pages: Miserable Failure, Snopes.com (Retrieved July 9, 2010); Official Google Webmaster Central Blog: A quick word about Googlebombs, Googlewebmastercentral.blogspot.com. January 25, 2007 (Retrieved July 9, 2010).

[60] Is CTR A Ranking Factor in Organic Results?, Search Engine Land, August 12, 2015 (Retrieved May 14, 2022), https://searchengineland.com/ctr-ranking-factor-227162

results gets noticed, an overcorrection towards the other end of the display spectrum may occur in order to punish and deter searchplace manipulation. Bing can exclude related search bubble suggestion buttons if harmful or violent content pops up. Amazon can curb its search results to specific content results if the suspicion exists that unrelated products are maliciously pegged to a competitor to bring down credibility or water down a specific big data cloud formation. Searchplaces permit owners of websites to disavow harmful and spamming backlinks pegged to websites. All these measures, however, require the awareness, attention, knowledge and pro-active engagement of the searchplace-discriminated victim.

To this day, however, behavioral black hat strategies that have become prominent to be used competitively in searchplaces, such as Google, Yahoo or Bing, are not well-regulated by governmental or governance efforts. The economic power dynamics and legal situation around competitive search engine results manipulation and its wider implications for entire affected domains are still not sufficiently covered in the academic and practitioners' literature.

Studying the socio-economic market dynamics of SEDO could help in order to present potential legislative, regulatory and institutional remedies to curb harmful market, socio-economic and individual consequences of searchplace discrimination. To this day, the currently available remedies are inadequate to provide effective relief for victims of searchplace discrimination, demanding for advances in behavioral economics and responsible finance leadership to concern the topic.

6.4.2.2 The role of behavioral insights for searchplace ethicality

The subliminal influence of environments on human decision making is studied by behavioral economics. Behavioral economics started in the 1950s with a theoretical critique of the neoclassical economic rational choice model in an iconic entrance of psychological and sociological contents in economics.

From the 1970s on, human decision-making was then investigated in laboratory and field experiments. Empirically, behavioral economists found heuristics as quick human decision making standards that often deviate from pure rationality. Behavioral science systematically outlines human decision-making deviations from rationality in mental shortcut heuristics. Heuristics were perceived as failures in the North American Behavioral Economics School, while the European tradition saw human decision making as successful strategy to cope with an overly complex world.

In the last decade, the behavioral economics approach was applied to political context in a behavioral insights revolution. Behavioral insights showed how to use nudging and winking to help citizens to make rational choices. Over time, behavioral insights specialists developed a broad range of nudges and winks to curb the harmful consequences of human decision-making or improve human fitness to adapt to an uncertain environment and complex world. Some of these powerful nudges to benefit from life and economic markets were communicated openly, while other behavioral insights were more subliminal change strategies.

From around the turn of the millennium, behavioral economics turned to big data online settings in order to guide human choices by strategic manipulation.

Behavioral Economics and Finance Leadership demonstrated how economics can be employed for the greater societal good, also when it comes to digitalization and online searchplaces. Most recently, systemic heuristics and biases are captured in big data of online observations.

While the primary focus of behavioral economics is on how to use nudges and winks to make the world a better place in helping humans make wiser decisions; to this day less is written and known about strategic manipulation of online environments by deleting information, clogging online searchplace users with unnecessary misinformation or compromising individual search profiles with disinformation.

Most recently, leadership and followership directives on nudging in digitalized spaces emerged that appeal to scholars and policy makers interested in rational decision-making and the use of nudging and winking in the digital age. With the advent of digitalization and the COVID-19 pandemic digital shock, but also in light of the lurking 5G revolution tying success even stronger to navigating a mounting big data online jungle, the time is ripe for behavioral insights to turn a critical eye to searchplace discrimination and dedicate behavioral insights attention to self-determined internet user empowerment. Behavioral Economics and Finance Leadership can now be advanced by aiding in creating inclusive digital spaces and discrimination-free online searchplaces in individualized-sensitive search results and self-determined search results reputation control.

6.4.2.3 Regulating searchplaces

Future behavioral insights could advance to improve the societal welfare in online virtual worlds and alleviate online inequalities. Legal advancements should include clear guidelines and oversight of fraudulent use of the Internet in a competitive way that manipulates genuine algorithm results, for instance, via clickfarms or backlinks that curb or tilt search results in a particular way. Regulatory advancements could draw from the wisdom developed in behavioral economics on how environments can form opinions and manipulate choices. Those who face a searchplace discrimination disadvantage should be protected by legal regulation, technical support and rescue funds established by the industry to uphold to favor quality over unethicality in their profession.

When it comes to declogging search engine results and the malicious pegging of content to unrelated or harmful information to bring down SEO, technological regulatory remedies and user empowerment strategies could expand on the U.S. Digital Millennium Copyright Act (DMCA) that allows for the removal of copyrighted material if used without consent by the copyright holder. Regulators could build on and/or strengthen the DMCA as a remedy to erase the unlawful use of copyrighted material on harmful, outdated or useless spamming webpages that curb a proper big data cloud formation and push down favorable and/or valuable content search engine results. Informing the public about the DMCA in general information campaigns and educating about the occurrence of SEDO strategies, could raise critical awareness and sensitivity for searchplace-discriminated victims and empower copyright holders and especially authors to maintain a fair online competition position.

When it comes to Internet rebels inappropriately using flagging and report buttons to weaken or make quality content of competitors unavailable, regulatory quality control and legally-enforceable human control over what gets flagged and who manipulates search results strategically is recommended. The previously-successfully-advocated "Right to Delete," could be turned into a "Right to not be Deleted." The "right to delete" or "right to be forgotten" developed by Mayer-Schönberger and adopted by the Court of Justice of the European Union (CJEU) allows for individuals to order the deletion of private information and removal of private information from Internet searches and other directories. This right could be revertedly extended into a "right to not be deleted" to instigate better quality control of online flagging tools with attentive oversight of humans instead of automated algorithmic action that likely can easier be manipulated.[61] As a punishment for flagging abuse, overemphasizing of strategically-incorrectly removed content may also deter from using SEDO techniques. Like in fines and legal punishment, repetitive malicious wrongdoing could grant the harmed victim higher credibility and compensation status, for instance, in the rising uplifting overrepresentation of erroneously-flagged important content.

Awareness building for SEDO tactics used in competitive settings will hopefully empower users to respond to inappropriate automatic erasing of quality content by search engines. Like in previous stakeholder advocacy, consumer pressure may instigate search engine operators to create the necessary fast, easy, and efficient communication channels to curb harmful reliance on algorithms being prone to e-heuristics. Suspicious data compartmentalization should become tracked by search engine providers and easily-accessible tools to visualize content clouds should be offered to decentralize search engine results presentations. People should have a right to know about their data clouds and have an active stance on shaping content that is pegged to their image online.

6.4.2.4 Stakeholder support against searchplace discrimination

Professional groups should align to curb unprofessional online conduct and offer information in trainings how to detect online cyberbullying by competitors and colleagues. The issue is particularly important in higher education and academia where online visibility and reputational capital are guideposts in developing successful careers. Concerted reporting mechanisms could become the basis for a registry of incidents that would lay open certain pockets of harmful institutions, professional groups and/or colleagues that may strategically remove competitors' contents online. Professional representation and union groups could drive advocacy for better institutional support against searchplace discrimination. Collegiality trainings and whistleblower protection to empower victims are future advancements to curb negative SEDO. Professional groups are in particular called for bringing up the problem of searchplace discrimination as bringing down colleagues

---

[61] Viktor Mayer-Schönberger, Delete: The Virtue of Forgetting in the Digital Age (2009).

and muffing excellence and ambition by rebellious searchplace users degrade entire research communities and stop academic discourse-driven advancements in competitive fields. Scanning and protection are especially needed during critical career moments—such as tenure decisions or when job candidates are competing on the job market.

Future behavioral law and economics advancements could directly investigate the societal burden and disparate impact of derailed digitalization on particular groups that hinder fair competition. Legal scholars may spearhead an analysis how to detect disparate impacts online in regard to vulnerable populations of untenured or job seeking academic candidates on the job market. Behavioral economists may work on incentivization schemes for victims to speak up against cyberbullies and online searchplace platform providers to punish discrimination (e.g., in overemphasizing positive search results of bullied) in order to deter from online misconduct. Technical support may be concerned with creating a right infrastructure to track online SEO performance and especially monitor changes as SEDO often appears malleable and unnoticingly gradual changes. Behavioral insiders could work on how to alleviate biases in an uncertain online world that changes quickly based on algorithmic heuristic choices. Behavioral leadership trainings could work toward uplifting and empowering weaker societal segments and in particular vulnerable groups, such as job market candidates, in our online digital workplaces and online searchplaces.

Awareness building in affected stakeholder groups is key—especially in social groups and decision-making leaders, such as behavioral economists, who are prone to understanding the often-unnoticed impact of favorable and unfavorable environmental "nudges" guiding individuals' choices, oftentimes subliminally. Awareness-raising should be provided to hiring authorities to not discriminate against cyberbullied job market candidates.

Self-help groups should be cultivated that empower against cyberbullying and exchange ideas to combat searchplace discrimination. Strategy exchange could include self-help tactics to piggyback on negative SEDO and Googlewashing. For instance, when bullies use Googlewashing or clickfarms to play up dormant Internet social media profiles without content (e.g., an old unused Facebook, Twitter, Instagram, or YouTube account with no followers), waiting until searchplaces features the account prominently in searches is recommended and then playing in a lot of quality content information at once can be a powerful strategy to replace spam with content and deter further harmful clickfarm attraction.

After all, sensitivity for cyberbullying via misinformation, disinformation, and spamming could elevate professions to a more ethical ground and uphold focus on excellence and merit rather than breeding discreditation potential due to lacking human-algorithmic control.

6.4.2.5 Future advancements in behavioral e-law and economics

The age of digitalization opens gates for searchplace discrimination, which is hardly captured in legal regulation or workplace anti-discrimination laws. Online inclusion should be free from any form of discrimination—may it be direct and obvious or more discrete. Ethics of online inclusion built on behavioral law and economics insights. Human-artificial intelligence algorithm compatibility is expected to leverage into a key advancement in behavioral economics and finance leadership of the future.

Future advancements in behavioral economics and finance leadership may address this abyss of discriminatory actions online and find ways how to avert the negative implications of searchplace black hat strategies. Most recent law and economics developments of the future could become the basis for solving practical ethical dilemmas arising from the disparate impact of negative SEDO strategies.

As a long-term goals, improving the algorithm-human-interaction gap could aid in quality control over online contents and content removal. The strategic display of information but also the competitive infiltration of online search results with harmful content, nonsense or unrelated cues should become subject to scrutiny and academic discourse ignited over upholding ethics and merit-based anti-discrimination. Anti-discrimination measures of the future should start to integrate insights about searchplace strategies. Searchplace providers, such as search engines but also social media tools and career platforms, should help refine search results and work toward wiser and more harmonious human-algorithmic interactions.

Ethics of inclusion, law and economics analyses and interdisciplinary dialogue building but also human-artificial intelligence algorithm compatibility and checks-and-balances are expected to become key advancements in behavioral economics and finance leadership of the future to tackle searchplace discrimination. All these endeavors may lead to an interdisciplinary understanding and sound regulatory framework for online inequality alleviation that can set the course for a better online future in a fairer, more credible and inclusive digital world.

6.4.3 Long COVID Information Exchange

The COVID-19 healthcare crisis and pandemic emergency around the world has exacerbated governmental control of data for monitoring, tracking and preventive prediction purposes [15, 53, 127, 128, 132]. A growing body of contemporary findings reveals that an estimated 10–30% of those previously infected with COVID-19 face some kind of long-term health impact and/or chronic debilitation that in many cases comes and goes in waves [60].

These so-called COVID Long Haulers are estimated to account for up to 1.9 billion people worldwide after the end of the pandemic. Given the large number of possible COVID Long Haulers, it is certain that this health phenomenon will have an enormous impact on society, medicine, the economy, the law and governance of our world [15, 136]. Long Haulers will contribute to the ongoing digitalization revolution by taking advantage of real-time health status and environmental infection condition tracking. As digitalization allowed for remote learning, working and entertainment, the biggest deurbanization trend in US history emerged in the wake of COVID-19 [127, 128, 136].

Labor market shortages and the number of workers quitting skyrocketing these days will likely further amplify a digitalization revolution to replace human contact and low-skilled labor [127, 128]. AI, robotics and big data insights come in handy when filling gaps for Long Haulers, who often face waves of debilitating conditions [139]. Robotics aid on patient care and hygiene [117]. Big data analytics have already revealed ground-breaking COVID long haul insights that will likely lead the way forward to finding remedies for those in chronic pain [136]. Moreover, Long Haulers have already found themselves in online self-help groups for quick and unbureaucratic information exchange about an emerging societal phenomenon [136].

Accounting for the nature, size and scope of the tragedy of Long COVID creates the imperative to protect the most vulnerable populations online when sharing sensitive information about their health and wellbeing in social media forums online from being turned against them [15]. Nowadays COVID long-haul patients have become—more than ever before—citizen scientists that bundle decentralized information on their health status and potential remedies in order to inform the medical profession about newly emerging trends. The rise in medical self-help and mutual support will have profound implications for the regulation of the medical profession and will likely stretch the medical remedy spectrum and boost alternative medicine. Instant online exchange of sensitive information about one's health status makes citizen scientists particularly vulnerable in terms of their privacy and potentially susceptible to online marketing campaigns under medically impaired conditions. But also, the long-term impact of publicly disclosed sensitive information that is shared bit-by-bit online over time appears sensitive.

In the digital age, it is difficult to estimate what effects the piecemeal providing of private information will have over time, when, for example, personal health information disseminated in an internet forum is absorbed into large datasets [15]. If information is analyzed and displayed in relation to other individuals' performance, a combined dataset could open gates for discrimination and stigmatization [15]. In the online exchange of sensitive information about one's health status, COVID Long Haulers, who recently have been recognized as potentially disabled group, are also in particular vulnerable in terms of their privacy, potentially susceptible to online marketing campaigns under medically impaired conditions, but also because of their sensitive information being publicly disclosed online over time [160]. This online sharing of medical information raises important—but yet hardly described—concerns about privacy, susceptibility to misinformation and discrimination in the vastly unregulated online social media arena, which calls for urgent attention.

6.4.4 Conclusion

Ethics of inclusion, Law and Economics interdisciplinary dialogue building and human-artificial intelligence algorithm compatibility are expected to become key advancements in behavioral economics and finance leadership of the future. Ethics of Inclusion should offer a comparative approach to understand the most contemporary responsibility challenges of our time. The idea of fairness and social justice should be elucidated from an ethics perspective in our post-pandemic era. Ethics of Inclusion will embrace diversified potentials that overall empower society. Inclusion should be free from any form of discrimination—may it be direct and obvious or more discrete. The age of digitalization opens gates for searchplace discrimination, which is hardly captured in legal regulation or workplace anti-discrimination laws.

Future advancements in behavioral economics and finance leadership may address this abyss of discriminatory actions online and find ways how to avert the negative implications of searchplace black hat strategies. For instance, improving the algorithm-human-interaction gap could aid in quality control over online contents and content removal. The strategic display of information but also the competitive infiltration of online search results with nonsense or unrelated content should become subject to scrutiny and academic discourse over upholding ethics and merit-based anti-discrimination. Anti-discrimination measures of the future should start to integrate insights about searchplace strategies. Searchplace providers, such as search engines but also social media tools and career platforms should help refine search results and work towards wiser and more harmonious human-algorithmic interactions.

Most recent law and economics developments should become the basis for solving practical ethical dilemmas arising in the fair distribution of healthcare, financial aid, education, digitalization and carrying the burden of climate stabilization. In the distribution of COVID-19 relief, a system change may be accomplished when being guided by the interdisciplinary insights of 'Law and Economics' that in particular can address the disparate impact of the disease in order to derived targeted inequality alleviation strategies.

Future economic policy research may be inspired by legal expertise on the measurement of disparate impact, which could open up the black box of the neoclassical aggregate economic functions calculus as a measure of economic growth. The combined expertise of Law and Economics in their unique interaction may shed light on disparate impacts of disease, public health and economic relief measures. This may lead to an interdisciplinary framework for crisis alleviation through redistribution that in combination can set the course for a better future in a more compassionate and inclusive world.

References

1. Aghion, P., & Bolton, P. (1997). A theory of trickle-down growth and development. *Review of Economic Studies, 64*(2), 151–172.
2. Aghion, P., Jones, B. F., & Jones, C. I. (2017). *Artificial intelligence and economic growth* (National Bureau of Economic Research Working Paper). https://www.nber.org/chapters/c14015

3. Ali, N. S., & Benabou, R. (2017). *Image versus information: Changing societal norms and optimal privacy* (National Bureau of Economic Research Working Paper 22203). https://scholar.princeton.edu/sites/default/files/rbenabou/files/imagevsinfo-21_002.pdf
4. Altman, I. (1975). *The environment and social behavior: Privacy, personal space, territory, and crowding*. Brooks.
5. Alvarez, F., Buera, F., & Lucas, R. E. (2007). *Idea flows, economic growth, and trade* (National Bureau of Economic Research Working Paper). https://www.nber.org/papers/w19667
6. Aravanis, J. (2020, April 7). Five industries set to outperform due to COVID-19. *IBISWorld*. https://www.ibisworld.com/industry-insider/coronavirus-insights/five-industries-set-to-outperform-due-to-covid-19/
7. Arora, R. (2020, June 30). Which companies did well during the coronavirus pandemic? *Forbes*. https://www.forbes.com/sites/rohitarora/2020/06/30/which-companies-did-well-during-thecoronavirus-pandemic/#17fe6c9b7409
8. Aumann, R. J., & Kruskal, J. B. (1958). The coefficients in an allocation problem. *Naval Research Logistics Quarterly, 5*, 111–123.
9. Aumann, R. J., & Kruskal, J. B. (1959). Assigning quantitative values to qualitative factors in the naval electronics problem. *Naval Research Logistics Quarterly, 6*, 1–16.
10. Baldwin, R. (2017). *The great conversion: Information technology and the new globalization*. Harvard University Press.
11. Banerjee, A. (2008). Big answers for big questions: The presumption of growth policy. In Brookings Institute (Ed.), *What works in development? Thinking big and thinking small* (pp. 207–231). Brookings Institute.
12. Banerjee, A., & Duflo, E. (2005). Growth theory through the lens of development economics. In P. Aghion & S. Durlauf (Eds.), *Handbook of economic growth* (pp. 473–552). Elsevier.
13. Banerjee, A., & Moll, B. (2010). Why does misallocation persist? *American Economic Journal: Macroeconomics, 2*(1), 189–206.
14. Banerjee, A., & Newman, A. F. (1993). Occupational choice and the process of development. *Journal of Political Economy, 101*(2), 274–298.
15. Bariffi, F., & Puaschunder, J. M. (2021, October 17). Artificial intelligence and big data in the age of COVID-19. In *Proceedings of the 24th RAIS Conference on Social Sciences and Humanities*.
16. Barseghyan, L., & DiCecio, R. (2011). Entry costs, industry structure, and cross-country income and TFP differences. *Journal of Economic Theory, 146*(5), 1828–1851.
17. Bartik, A.W., Bertrand, M., Cullen, Z., Glaeser, E.L., Luca, M. & Stanton, Ch. (2020). How are small businesses adjusting to COVID-19? Early evidence from a survey. Cambridge, MA: National Bureau of Economic Research Working Paper No. 26989.
18. Beerbaum, D., & Puaschunder, J. M. (2018). A behavioral economics approach to digitalisation: The case of a principles-based taxonomy. In *Proceedings of the 10th International RAIS Conference on Social Sciences and Humanities* (Vol. 211). Published by Atlantis Press. Part of Series: ASSEHR. ISSN: 2352-5398.
19. Beerbaum, D., & Puaschunder, J. M. (2019a). A behavioral economics approach to digitalization: The case of a principles-based taxonomy. In J. M. Puaschunder (Ed.), *Intergenerational governance and leadership in the corporate world* (pp. 107–122). Idea Group.
20. Beerbaum, D., & Puaschunder, J. M. (2019b). A behavioral economics approach to digitalization: The case of a principles-based taxonomy. In *Proceedings of the 9th International RAIS Conference on Social Sciences and Humanities Organized by Research Association for Interdisciplinary Studies (RAIS) at The Erdman Center at Princeton University*, Princeton, NJ, August 22–23, 2018.
21. Beerbaum, D., & Puaschunder, J. M. (2019c). A behavioral approach to irrational exuberances: An artificial intelligence roboethics taxonomy. *Scientia Moralitas: International Journal of Multidisciplinary Research, 4*(1), 1–30.
22. Besley, T., & Prat, A. (2006). Handcuffs for the grabbing hand? Media capture and government accountability. *American Economic Review, 96*(3), 720–736.

23. Brenner, R. (2002). *The boom and the bubble: The US in the world economy*. Verso.
24. Brenner, R. (2006). *The economics of global turbulence: The advanced capitalist economies from long boom to long downturn, 1945–2005*. Verso.
25. Buera, F. J., & Kaboski, J. P. (2012). The rise of the service economy. *American Economic Review, 102*(6), 2540–2569.
26. Buera, F. J., & Shin, Y. (2011). Finance and development: A tale of two sectors. *American Economic Review, 101,* 1964–2002.
27. Centeno, M. A., Cinlar, E., Cloud, D., Creager, A. N., DiMaggio, P. J., Dixit, A. K., et al. (2013a, April). *Global systemic risk* (Unpublished manuscript for research community). Princeton Institute for International and Regional Studies, Princeton University.
28. Centeno, M. A., Creager, A. N., Elga, A., Felton, E., Katz, S. N., Massey, W. A., & Shapiro, J. N. (2013b). *Global systemic risk: Proposal for a research community* (Unpublished Working Paper). Princeton University Institute for International and Regional Studies.
29. Centeno, M. A., & Tham, A. (2012). *The emergence of risk in the global system* (Princeton University Working Paper).
30. Centeno, M. A., Cinlar, E., Cloud, D., Creager, A. N., DiMaggio, P. J., Dixit, A. K., Elga, A. N., Felten, E. W., James, H., Katz, St. N.,Keohane, R. O., Leonard, Th. C., Massey, W. A., Mian, A. R., Mian, Z., Oppenheimer, M., Shafir, E. & Shapiro, J. N. (2013). Globalsystemic risk. Unpublished manuscript for research community. Princeton Institute for International and Regional Studies, Princeton University, April 2013.
31. Clancy, E. (1998). The tragedy of the global commons. *Indiana Journal of Global Legal Studies, 5*(2), 601–619.
32. Corlatean, T. (2020, June). Risk, discrimination and opportunities for education during the times of COVID-19 pandemic. In *Proceedings of the 17th RAIS Research Association for Interdisciplinary Sciences Conference on Social Sciences and Humanities* (pp. 37–46). http://rais.education/wp-content/uploads/2020/06/004TC.pdf
33. Crémer, J., Garicano, L., & Prat, A. (2007). Language and the theory of the firm. *The Quarterly Journal of Economics, 122*(1), 373–407.
34. Delli Carpini, M. X., & Keeter, S. (1989). *What Americans know about politics and why it matters*. Yale University Press.
35. Dillon, M., & Garland, L. (2005). *Ancient Rome: From the early republic to the assassination of Julius Caesar*. Routledge.
36. Dodd, D. (2020, June 22). COVID 19's corporate casualties. *Financial Times*. https://www.ft.com/content/eb6efc36-bf99-4086-a98a-7d121738b4b4
37. Dua, A., Ellingrud, K., Mahajan, D. & Silberg, J. (2020). Which small businesses are most vulnerable to COVID-19: And when, McKinsey & Co., June 18, 2020. Retrieved at https://www.mckinsey.com/featured-insights/americas/which-small-businesses-are-mostvu lnerable-to-covid-19-and-when
38. Duarte, M., & Restuccia, D. (2006). The productivity of nations. *Federal Reserve Bank Richmond Economic Quarterly, 92*(3), 195–223.
39. Echevarria, C. (1997). Changes in sectoral composition associated with economic growth. *International Economic Review, 38*(2), 431–452.
40. EcoWellness Group. (2020, July 14–15). *Salzburg declaration: Interdisciplinary conference on 'system change?! The chance of transformation of the healthcare system: Analysis and chances of the coronavirus crisis*. https://www.oekowellness.de/laenderuebergreifende-kon zerenz-zum-thema-system-change-die-chance-der-transformation-des-gesundheitswesens-14-07-2020/
41. EcoWellness Group. (2021, July 14–15). *Salzburg European declaration from the gasteinertal: Interdisciplinary conference on 'system change?! The chance of transformation of the healthcare system*. Retrieved August 23, 2021, from https://www.oekowellness.de/wp-con tent/uploads/2021/07/Final-Stand-5.7.-2021.07.04_Programm-14.7.-und-15.07.2021-2.pdf
42. Elliott, C. (2021, February 5). When will it be safe to cruise again? These signs that will help you decide when to sail. *USA Today*. https://www.usatoday.com/story/travel/advice/2021/02/05/covid-when-will-it-be-safe-to-cruise/4386762001

43. Etzioni, A. (2012). The privacy merchants: What is to be done? *The Journal of Constitutional Law, 14*(4), 929–951.
44. Fishburn, P. C. (1968). Utility theory. *Management Science, 14*(5), 335–378.
45. Flaherty, D. (1989). *Protecting privacy in surveillance societies: The federal republic of Germany, Sweden, France, Canada, and the United States*. The University of North Carolina Press.
46. Foster, A. D., & Rosenzweig, M. R. (2010). *Microeconomics of technology adoption* (Yale University Economic Growth Center Discussion Paper 984).
47. Foster, V. (2020, June 21). Is eating meat from meatpacking plants with COVID-19 coronavirus outbreaks safe? *Forbes*. https://www.forbes.com/sites/victoriaforster/2020/06/21/is-eating-meatfrom-meatpacking-plants-with-covid-19-coronavirus-outbreaks-safe/?sh=5d2d5bcb7089
48. Fowler, R., Hodge, B., Kress, G., & Trew, T. (1979). *Language and control*. Routledge.
49. Galor, O., & Moav, O. (2000). Ability biased technological transition, wage inequality and growth. *Quarterly Journal of Economics, 115*, 469–498.
50. Galor, O., & Moav, O. (2004). From physical to human capital accumulation: Inequality in the process of development. *Review of Economic Studies, 71*, 1001–1026.
51. Gaudeul, A., & Giannetti, C. (2017). The effect of privacy concerns and social network formation. *Journal of Economic Behavior & Organization, 141*, 233–253.
52. Gelter, M. (2021). Is economic nationalism in corporate governance always threat? *Ohio State Business Law Journal, 16*(1), 1–36.
53. Gelter, M., & Puaschunder, J. M. (2021). COVID-19 and comparative corporate governance. *Journal of Corporation Law, 46*, 557–629.
54. Gelter, M. & Puaschunder, J.M. (2021). COVID-19 and comparative corporate governance. Journal of Corporation Law, 46, 3, 557- 629.
55. Goldberg, K., & Pavcnik, N. (2007). The distributional effects of globalization in developing countries. *Journal of Economic Literature, 45*(1), 39–82.
56. Greenberg, A., & Collins, S. (1966). Paired comparison taste tests: Some food for thought. *Journal of Marketing Research, 3*, 76–80.
57. Greenwood, J., & Jovanovic, B. (1990). Financial development, growth, and the distribution of income. *Journal of Political Economy, 98*(5), part 1, 1076–1107.
58. Hagey, K., Alpert, L. I., & Serkez, Y. (2019, May 4). In news industry, a stark divide between haves and have-nots: Local newspapers are failing to make the digital transition larger players did and are in danger of vanishing. *The Wall Street Journal*.
59. Harris, W. V. (2000). Trade. In *The Cambridge ancient history: The high empire A.D. 70–192* (p. 11). Cambridge University Press.
60. Hart, R. (2021, July 15). Long covid has over 200 symptoms and leaves 1 in 5 unable to work, study finds. *Forbes*. https://www.forbes.com/sites/roberthart/2021/07/15/long-covid-has-over-200-symptoms-and-leaves-1-in-5-unable-to-work-study-finds/?sh=7f71338e5eb2
61. Held, D., & McGrew, A. G. (2007). *A new world economic order? Global markets and state power: Beyond globalization/anti-globalization: Beyond the great divide*. Polity.
62. International Monetary Fund. (2020). *Policy responses to COVID-19*. Retrieved August 13, 2020, from https://www.imf.org/en/Topics/imf-andcovid19/Policy-Responses-to-COVID-19
63. Iyengar, S., & Kinder, D. R. (1987). *News that matter*. University of Chicago Press.
64. Jacoby, H. G. (2008). Food prices, wages, and welfare in rural India. *Economic Inquiry, 54*(1), 159–176.
65. Jevons, W. T. (1871). *The theory of political economy*. Macmillan. http://www.econlib.org/library/YPDBooks/Jevons/jvnPECover.html
66. Johnson, D. (2009). *Ethical theory and business*. Pearson Prentice Hall.
67. Jones, L., Palumbo, D., & Brown, D. (2020, June 30). Coronavirus: A visual guide to the economic impact. *BBC News*. https://www.bbc.com/news/business-51706225
68. Kahneman, D., & Tversky, A. (1979). Prospect theory: An analysis of decision under risk. *Econometrica, 47*(2), 263–291.

69. Kirchler, E. M. (2007). *The economic psychology of tax behaviour*. Cambridge University Press.
70. Kirchler, E. M. (2011). *Wirtschaftspsychologie: Individuen, Gruppen, Märkte, Staat*. Hogrefe.
71. Klenow, P. (2008). Discussion of 'Big answers for big questions: The presumption of growth policy' by A.V. Banerjee. In Brookings Institute (Ed.), *What works in development? Thinking big and thinking small* (pp. 207–231). Brookings Institute.
72. Kosinski, M., Bachrach, Y., Kohli, P., Stillwell, D. J., & Graepel, T. (2014). Manifestations of user personality in website choice and behavior on online social networks. *Machine Learning Journal, 95*, 357–380.
73. Kosinski, M., Stillwell, D., & Graepel, T. (2013). Private traits and attributes are predicable from digital records of human behavior. *Proceedings of the National Academy of Sciences, 110*(15), 5802–5805.
74. Kremer, M., Rao, G., & Schilbach, F. (2019). Behavioral development economics. In D. Bernheim, St. DellaVigna, & D. Laibson (Eds.), *Handbook of behavioral economics: Foundations and applications* (pp. 345–458). Elsevier.
75. Kumar, N., & Haydon, D. (2020, April 7). Industries most and least impacted by COVID-19 from a probability of default perspective: March 2020 update. *S&P Global Market Intelligence*. https://www.spglobal.com/marketintelligence/en/news-insights/blog/industries-most-and-least-impacted-by-covid-19-from-a-probability-of-default-perspective-march-2020-update
76. Kwak, J. (2020). The end of small business, The Washington Post, July 9, 2020, https://www.washingtonpost.com/outlook/2020/07/09/after-covid-19-giant-corporations-chains-may-be-only-ones-left/?arc404=true
77. Lessig, L. (2006). *Code: Version 2.0*. Basic.
78. Lin, P., Abney, K., & Bekey, G. A. (2012). *Robot ethics: The ethical and social implications of robotics*. The MIT Press.
79. Lucas, R. E. (2004, May). The industrial revolution: Past and future. In *Annual report of the Federal Reserve Bank of Minneapolis* (pp. 5–20).
80. Lucas, R. E., & Moll, B. (2014). Knowledge growth and the allocation of time. *Journal of Political Economy, 122*(1), 1–51.
81. MacDonald, F. (2016, June 23). Harvard scientists think they've pinpointed the physical source of consciousness. *Science Alert*. http://www.sciencealert.com/harvard-scientists-think-they-ve-pinpointed-the-neural-source-of-consciousness
82. Marquardt, R., Makens, J., & Larzelere, H. (1965). Measuring the utility added by branding and grading. *Journal of Marketing Research, 2*, 45–50.
83. Marra, W., & McNeil, S. (2013). Understanding "the loop": Regulating the next generation of war machines. *Harvard Journal of Law & Public Policy, 36*, 1139–1187.
84. Matsuyama, K. (2008). Structural change. In S. Durlauf & L. E. Blume (Eds.), *The new Palgrave dictionary of economics* (pp. 410–420). Palgrave-Macmillan.
85. Matsuyama, K. (2011). Imperfect credit markets, household wealth distribution, and development. *Annual Review of Economics, 3*, 339–362.
86. Mauss, M. (1979). A category of the human mind: The notion of the person, the notion of 'self'. In M. Mauss (Ed.), *Sociology and Psychology* (pp. 81–103). Routledge.
87. Mayer-Schönberger, V. (2009). *Delete: The virtue of forgetting in the digital age*. Princeton University Press.
88. Meghdari, A., & Alemi, M. (2018). Recent advances in social & cognitive robotics and imminent ethical challenges. In *Proceedings of the 10th International RAIS Conference on Social Sciences and Humanities Organized by Research Association for Interdisciplinary Studies (RAIS) at The Erdman Center at Princeton University*, Princeton, NJ. The Scientific Press.
89. Milanovic, B. (2013). Global income inequality in numbers: In history and now. *Global Policy, 4*(2), 198–208.
90. Mohamed, A. M. I. (2016). *Globalization and new international public works agreements in developing countries: An analytical perspective*. Routledge.

91. Moll, B. (2014). Productivity losses from financial frictions: Can self-financing undo capital misallocations? *American Economic Review, 104*(10), 3186–3221.
92. Mookherjee, D., & Napel, S. (2007). Intergenerational mobility and macroeconomic history dependence. *Journal of Economic Theory, 137*(1), 49–78.
93. Mookherjee, D., & Ray, D. (2003). Persistent inequality. *Review of Economic Studies, 70,* 369–393.
94. Mumford, A. (2001). *Taxing culture*. Ashgate.
95. Murphy, K. M., Riddell, W. C., & Romer, P. M. (1998). *Wages, skills, and technology in the United States and Canada* (National Bureau of Economic Research (NBER) Working Paper 6638). https://www.nber.org/papers/w6638.pdf
96. Neuman, W., Russell, M. R. J., & Crigler, A. N. (1992). *Common knowledge*. Chicago University Press.
97. Norris, P., & Sanders, D. (2003). Message or medium? Campaign learning during the 2001 British General Election. *Political Communication, 20,* 233–262.
98. Olken, B. A. (2009). Do television and radio destroy social capital? Evidence from Indonesian villages. *American Economic Journal: Applied Economics, 1*(4), 1–33.
99. Orwell, G. (1949). *1984*. Harcourt Brace.
100. Oulasvirta, A., Suomalainen, T., Hamari, J., Lampinen, A., & Karvonen, K. (2014). Transparency of intentions decreases privacy concerns in ubiquitous surveillance. *Cyberpsychology, Behavior, and Social Networking, 17*(10), 633–638.
101. Parente, St., & Prescott, E. (1993). Changes in the wealth of nations. *Quarterly Review of Economics, 17*(2), 3–16.
102. Piketty, T. (2014). *Capital in the twenty-first century*. Harvard University Press.
103. Posner, R. A. (1981). The economic of privacy. *The American Economic Review, 71*(2), 405–409.
104. Prat, A. (2017). *Lecture notes 'industrial organization'*. Columbia University.
105. Price, L. (2020). Impact of COVID-19 on small businesses: Where is it worst?, Small Business Trends, April 29, 2020. Retrieved at https://smallbiztrends.com/2020/04/impact-of-coronavirus-on-small-businesses.html
106. Prat, A., & Strömberg, D. (2005, April). *Commercial television and voter information* (CEPR Discussion Paper 4989).
107. Prat, A., & Strömberg, D. (2013). *The political economy of mass media* (Columbia University Working Paper). http://www.columbia.edu/~ap3116/papers/mediasurvey11.pdf
108. Profita, S. (2019). *Slowbalization and its risks* (Columbia University Working Paper). Columbia University.
109. Puaschunder, J. M. (2017a). The nudging divide in the digital big data era. *International Journal of Research in Business, Economics and Management, 4*(11–12), 49–53.
110. Puaschunder, J. M. (2017b). We—Today's and tomorrow's—People of the United World: Rethinking capitalism for intergenerational justice in the fin-de-millénaire. *Corporate Governance and Sustainability Review, 1*(2), 30–35.
111. Puaschunder, J. M. (2018a). *A utility theory of privacy and information sharing* (Social Science Research Network Working Paper). https://papers.ssrn.com/sol3/papers.cfm?abstract_id=3197744
112. Puaschunder, J. M. (2018b). *Artificial intelligence ethik* (Social Science Research Network paper). https://papers.ssrn.com/sol3/papers.cfm?abstract_id=3137926
113. Puaschunder, J. M. (2018c). *Artificial intelligence evolution: On the virtue of killing in the artificial age* (Social Science Research Network Working Paper). https://papers.ssrn.com/sol3/papers.cfm?abstract_id=3247401
114. Puaschunder, J. M. (2018d). Gifts without borders: Intergenerational glue connecting over distance and time as pure international development in the age of migration. In J. M. Puaschunder (Ed.), *Intergenerational responsibility in the 21st century* (pp. 143–174). Vernon.
115. Puaschunder, J. M. (2019a). An inquiry into the nature and causes of Climate Wealth of Nations: What temperature finance gravitates towards? Sketching a climate-finance nexus and

outlook on climate change-induced finance prospects. *Archives of Business Research, 7*(3), 183–217.

116. Puaschunder, J. M. (2019b). Artificial intelligence evolution: On the virtue of killing in the artificial age. *Journal of Sociology, 3*(1), 10–29.
117. Puaschunder, J. M. (2019c). Artificial intelligence market disruption. In *Proceedings of the International RAIS Conference on Social Sciences and Humanities Organized by Research Association for Interdisciplinary Studies (RAIS) at Johns Hopkins University*, Montgomery County Campus, Rockville, MD, June 10–11 (pp. 1–8).
118. Puaschunder, J. M. (2019d). Dignity and utility of privacy and information sharing in the digital big data age. *International Journal of Commerce and Management Research, 5*(4), 62–70.
119. Puaschunder, J. M. (2019e). On artificial intelligence's razor's edge: On the future of democracy and society in the artificial age. *Journal of Economics and Business, 2*(1), 100–119.
120. Puaschunder, J. M. (2019f). On artificial intelligence's razor's edge: On the future of democracy and society in the artificial age. *Scientia Moralitas: International Journal of Multidisciplinary Research, 4*(1), 51–72.
121. Puaschunder, J. M. (2019g). *The legal and international situation of AI, robotics and big data with attention to healthcare* (Report on behalf of the European Liberal Forum of the European Parliament). European Union.
122. Puaschunder, J. M. (2019h). Towards a utility theory of privacy and information sharing: The introduction of hyper-hyperbolic discounting. *International Journal of Strategic Information Technology and Applications (IJSITA), 10*(1), 1–22.
123. Puaschunder, J. M. (2019i). Towards a utility theory of privacy and information sharing and the introduction of hyper-hyperbolic discounting in the digital big data age. In E. Idemudia (Ed.), *Handbook of research on social and organizational dynamics in the digital era* (pp. 157–200). IGI.
124. Puaschunder, J. M. (2019j). Towards a utility theory of privacy and information sharing and the introduction of hyper-hyperbolic discounting in the digital big data age. *Research Association for Interdisciplinary Studies Collective Volume: Economic Science, 2019*, 4–46.
125. Puaschunder, J. M. (2019k). *Revising growth theory in the artificial age: Putty and clay labor* (Princeton University Working Paper).
126. Puaschunder, J. M. (2020a, September 28–29). Economic growth in times of pandemics. In *Proceedings of the ConScienS Conference on Science & Society: Pandemics and Their Impact on Society* (pp. 1–9).
127. Puaschunder, J. M. (2020b, October 30). The Green New Deal: Historical foundations, economic fundamentals and implementation strategies. *The FinReg Blog: Global Financial Markets Center Duke University School of Law Blog*. https://sites.law.duke.edu/thefinregblog/author/julia-m-puaschunder/
128. Puaschunder, J. M. (2020c). *Governance and climate justice: Global south and developing nations*. Palgrave Macmillan, Springer Nature.
129. Puaschunder, J. M. (2021a). Alleviating COVID-19 inequality. In *ConScienS Conference Proceedings* (pp. 185–190).
130. Puaschunder, J. M. (2021b). Focusing covid-19 bailout and recovery. *Ohio State Business Law Journal, 16*(1), 91–148.
131. Puaschunder, J. M. (2021c, October 17). Law & economics. In *Proceedings of the 24rd Research Association for Interdisciplinary Studies (RAIS) Conference*.
132. Puaschunder, J. M. (2022a). *Ethics of inclusion: The cases of health, economics, education, digitalization and the environment in the post-COVID-19 era*. Ethics International Press.
133. Puaschunder, J. M. (2022b). The future of artificial intelligence in international healthcare: An index. In H. Nagl-Docekal & W. Zacharasiewicz (Eds.), *Proceedings of the Transatlantic Dialog Conference on Artificial Intelligence and Human Enhancement: Affirmative and Critical Approaches in the Humanities from Both Sides of the Atlantic Conference* (pp. 181–206). Austrian Academy of Sciences, Austria, European Union, De Gruyter.

134. Puaschunder, J. M. (forthcoming). *On the collective soul of booms and busts*. https://papers.ssrn.com/sol3/papers.cfm?abstract_id=2799646
135. Puaschunder, J. M., & Beerbaum, D. (2020, September 28–29). Healthcare inequality in the digital 21st century: The case for a mandate for equal access to quality medicine for all. In *Proceedings of the 1st Unequal World Conference: On Human Development*, United Nations.
136. Puaschunder, J. M., & Gelter, M. (2022). The law, economics and governance of generation COVID-19 long-haul. *Indiana Health Law Review, 19*(1), 47–125.
137. Puaschunder, J. M., Gelter, M., & Sharma, S. (2020a, September 28–29). Alleviating an unequal COVID-19 world: Globally digital and productively healthy. In *Proceedings of the 1st Unequal World Conference: On Human Development*, United Nations.
138. Puaschunder, J. M., Gelter, M., & Sharma, S. (2020b, August 17–18). COVID-19-shock: Sociotechnological, legal, corporate, economic and governance changes. In *Proceedings of the 18th Interdisciplinary Research Association for Interdisciplinary Studies (RAIS) Conference*.
139. Puaschunder, J.M. & Gelter, M. (2022). The law, economics, and governance of generation COVID-19 long-haul. Indiana Health Law Review, 19(1), 47-125. DOI: https://doi.org/10.18060/26085
140. Putnam, R. D. (2000). *Bowling alone: The collapse and revival of American community*. Simon & Schuster.
141. Rajan, R. G., & Zingales, L. (1998). Financial dependence and growth. *American Economic Review, 88*(3), 559–586.
142. Read, D. R. (1964). A quantitative approach to the comparative assessment of taste quality in the confectionery industry. *Biometrics, 20,* 143–155.
143. Restuccia, D., & Rogerson, R. (2017). The causes and costs of misallocation. *Journal of Economic Perspectives, 31*(3), 151–174.
144. Restuccia, D., & Urrutia, C. (2001). Relative prices and investment rates. *Journal of Monetary Economics, 47*(1), 93–121.
145. Rodrik, D. (2006). Goodbye Washington consensus, hello Washington confusion? A review of the World Bank's economic growth in the 1990s: Learning from a decade of reform. *Journal of Economic Literature, 44*(4), 973–987.
146. Russell, St., & Norvig, P. (1995). *Artificial intelligence a modern approach*. Simon & Schuster.
147. Saffari, E., Meghdari, A., Vazirnezhad, B., & Alemi, M. (2015). Ava (a social robot): Design and performance of a robotic hearing apparatus. In *LNCS: Social Robotics* (Vol. 9388, pp. 440–450). Springer.
148. Schumpeter, J. A. (1942/1975). *Capitalism, socialism and democracy*. Harper.
149. Shannon, C. E. (1948). A mathematical theory of communication. *The Bell System Technical Journal, 27,* 379–423.
150. Snyder, J. M., & Strömberg, D. (2010). Press coverage and political accountability. *Journal of Political Economy, 118*(2), 355–408.
151. Solove, D. J. (2008). *Understanding privacy*. Harvard University Press.
152. Solum, L. (1992). Legal personhood for artificial intelligences. *North Carolina Law Review, 70*(4), 1231–1287.
153. Stafford, J. E. (1966). Effects of group influences on consumer brand preferences. *Journal of Marketing Research, 3,* 68–75.
154. Stiglitz, J. E. (1998). The private uses of public interests: Incentives and institutions. *Journal of Economic Perspectives, 12,* 3–22.
155. Stillson, P. (1954). A method for defect evaluation. *Industrial Quality Control, 11,* 9–12.
156. Stroebe, W., & Frey, B. S. (1982). Self-interest and collective action: The economics and psychology of public goods. *British Journal of Social Psychology, 21*(2), 121–137.
157. Suzuki, G. (1957). Procurement and allocation of naval electronic equipments. *Naval Research Logistics Quarterly, 4,* 1–7.
158. *The Economist*. (2019, January 26). *The steam has gone out of globalisation: Slowbalisation* (pp. 17–20).

159. *The Economist*. (2020). *The pandemic has caused the world's economies to diverge*. https://www.economist.com/leaders/2020/10/08/the-pandemic-has-caused-the-worlds-economies-to-diverge
160. *The White House of The United States of America*. (2021). https://www.whitehouse.gov/briefing-room/legislation/2021/01/20/president-biden-announces-american-rescue-plan/
161. The World Economic Forum. (2021). *The World Economic Forum great reset*. https://www.weforum.org/great-reset/
162. Themistoklis, T. (2018). Artificial intelligence as global commons and the "international law supremacy" principle. In *Proceedings of the 10th International RAIS Conference on Social Sciences and Humanities Organized by Research Association for Interdisciplinary Studies (RAIS) at The Erdman Center at Princeton University*, Princeton, NJ. The Scientific Press.
163. Townsend, R. M., & Ueda, K. (2006). Financial deepening, inequality, and growth: A model-based quantitative evaluation. *Review of Economic Studies, 73*(1), 251–293.
164. Tybout, J. (2000). Manufacturing firms in developing countries: How well do they do, and why? *Journal of Economic Literature, 38*(1), 11–44.
165. Ventura, J. (1997). Growth and interdependence. *The Quarterly Journal of Economics, 112*(1), 57–84.
166. Vlassopoulos, K. (2009). *Politics antiquity and its legacy*. Oxford University Press.
167. Warren, S., & Brandeis, L. (1890). The right to privacy. *Harvard Law Review, 4*(5), 193–220.
168. Warwick, K., & Shah, H. (2014). How good robots will enhance human life. In K. Tchon & W. W. Gasparski (Eds.), *Treatise on good robots edition: Praxology: The international annual of practical philosophy and methodology* (pp. 3–19). Transaction.
169. Witzany, G. (2012). *Biocommunication of fungi*. Springer.
170. World Bank 2008 Report. (2008). *Technology diffusion in the developing world: Global economic prospects report*. World Bank.
171. World Bank Group Migration and Development Brief 26. (2016, April). *Migration and remittances: Recent development and outlook*. International Bank for Reconstruction and Development, World Bank Group.
172. World Development Report. (2015). *Mind, society and behavior*. World Bank.
173. Wu, A. H. (2018). Gendered language on the economics job market rumors forum. *American Economic Association Papers and Proceedings, 108*, 175–179.

Discussion

7

The book presented a first theoretical introduction of a utility theory of information sharing and privacy. Potential limitations are that some communication may not be integrated into the framework, such as nonverbal communication or emotional responses. In general, information exchange is very heterogeneous and vast international differences are assumed to exist. In addition, in what time online communication and under what circumstances decisions regarding communication and privacy are made remains a completely undiscussed topic.

As a next research step, a stringent hypothesis testing of the presented problem is recommended. For instance, future research projects featuring a multi-methodological approach will help gain invaluable information about the actual performance and behavior regarding information sharing and privacy upholding. Interaction of individuals on social media should be scrutinized in order to derive real-world relevant economic insights for legal and policy-making purposes alongside advancing an upcoming scientific field.

Following empirical investigations should employ a critical survey of the intersection of analytic and behavioral perspectives to decision-making in information sharing. Literature discussion featuring a critical analysis of how to improve e-literacy should be coupled with e-education and enhancement of e-ethicality. Research should be directed toward a critical analysis of the application of behavioral economics on hyper-hyperbolic discounting in the digital age.

In the behavioral economics domain, both approaches, studying the negative implications of information sharing and decision-making to uphold privacy but also finding ways how to train new media users' wiser decisions should be explored. Negative aspects of Internet representations, such as the targeted unfair manipulation of search algorithms and online social media information turned against individuals, must be addressed in order to make the collective experience of the Internet more accurate and fairer. Interdisciplinary viewpoints and multi-method research approaches should be covered in the heterodox economics readings but also in a variety of independent individual research projects. Research support

J. Puaschunder, *Advances in Behavioral Economics and Finance Leadership*, Contributions to Economics, https://doi.org/10.1007/978-3-031-15710-3_7

and guidance should be targeted at nurturing interdisciplinary research interests on privacy and information sharing in the fields of behavioral economics and public affairs.

More concretely, future studies should define the value that data has to individuals and data sovereignty in the international context. When people share information, they should be informed to consider what the benefit and value of information sharing are for them and what the benefit for social media industrialists-capitalists is. The sovereignty of data and the human dignity of privacy should become debated as civic virtual virtue in the twenty-first century. Individuals should be informed that sharing data is a personal security risk, if considered to be asked for social media information upon entry of a country.

Future studies should describe what companies and institutions constitute the complex system that helps establishing the nudgital society and the influence that social media has. The implicit underlying social structure of the nudgital society based on a complicated information gathering machinery should become subject to scrutiny and how, in particular, the nudgital class division is supported by a comprehensive social network data processing method. How social media advertising space can be used to specialize in targeted propaganda and misleading information to nudge the populace in an unfavorable way should be unraveled.

Search function algorithms, but also the outsourcing of criminal and dishonest activities to providers from less regulated parts of the Internet appear as a new interpretation of Internet security and consumer protection online.

In the public domain, the role of politicians in using different communication channels and tools to manipulate the population through targeted communication and novel online media platforms should be examined. The study of heuristics could elucidate that heuristics played in elections as they made people less likely to vote logically. This would be key as it would help explain how people chose to vote, and why they do not always make the most logical choice when voting. This line of research could help to more accurately promote future elections' candidates, how to better predict election outcomes, and how to improve democracy.

In voting behavior and the candidates' media communication gains and the value of detailed market information was never before as tied to social online media presence as it is today. The current research in this area lacks empirical evidence, demanding for further investigations on how nudges can directly impact individual's choices and new media has become a governance manipulation tool. What social instruments are employed on social media and what prospects data processing has in light of privacy infringement lawsuits should be uncovered. How social media is utilized to create more favorable social personas for political candidates should be explored. How Internet online presences allow gainingo gain as much attraction as possible for the presence of political candidates is another question of concern. Another area of interest is how selective representations influence the voting population and what institutions and online providers are enabling repetitiveness and selectivity. How gathered individual information is used to parse data to manipulate social Internet behavior and subsequent action is another topic to be investigated.

Future research goals will include determining what this means for the future political landscape and how Internet users should react to political appearances online. Information should be gathered on how we choose what media to watch and if political views play a role in media selection and retention. Does distrust in the media further political polarization and partisanship needs to be clarified. Future studies should also look into the relationship between individual's political ideologies and how they use and interact on social media, especially with a focus on the concept of fake news and alternative facts. Where do these trends come from and who is more susceptible to these negative impacts of the digital society? Has social media become a tool to further polarize political camps is needed to be asked.

Future investigations could also use decision anomalies to explain the outcomes of presidential elections and leadership communication on social media, to show that knowledge of heuristics plays a key role in massively influencing the likelihood of voting logically and making informed decisions. This would be crucial as it would help explain how people vote and why social media can influence them when doing so. This line of research could help to better target future political candidates for elections and better inform the masses in order to make election results more democratic and predictable, and generally improve empowered democracy. All these endeavors will help outline the existence of social media's influence in governance and data processing to aid political campaigning in order to derive inferences about democracy and political ethicality in the digital age.

In addition, nudging through means of visual merchandising, marketing, and advertising should be captured in order to uphold ethical standards in social media. Nudging's role in selling products, maximizing profits but also creating political trends should be uncovered. While there is knowledge of the visual merchandising in stores and window displays, little appears to be known how online appearances can nudge people into making certain choices. In particular, the familiarity heuristic, anchoring, and the availability heuristics may play a role in implicitly guiding people's choices and discreetly persuading consumers and the populace. Not to mention advancements of online shopping integrity and e-commerce ethics, the prospective insights gained will aid upholding moral standards in economic marketplaces and hopefully improve democratic outcomes of voting choices.

Contemporary studies could also address if the age of instant messaging has led to a loss of knowledge in information sharing. Future research should also investigate how search engines can be manipulated to make favorable sources more relevant and how artificial intelligence and social networks can become dangerous data manipulation means. Future research should also examine how search engines can be manipulated to make sources more or less relevant, and how AI and social networks can be turned into dangerous data and image manipulation tools. This can be observed, for example, against competitors in competitive situations when it comes to Search Engine Optimization blackhat strategies, which manipulate online content's relevance. The overinflating unfavorable or irrelevant information of search engine results of competitors and the strategic cloaking of the Internet appearance of competitors in Googlewashing or the piggybacking on other's content all account for most novel unfair online market behavior that has hardly been described so far.

The role of data processing companies may be studied in relation to the idea of data monopoly advantages—hence situations in which data processing companies may utilize data flows for their own purposes to support sponsored causes or their own ideals. Due to the specific time period of the digital age no extrapolations to past time periods is possible but the results appear useful in determining future behaviors. Situations in which data processing companies use data streams for their own purposes to support sponsored activities or their own ideals appear ethically questionable.

How social media tools nudge people to not give everything at once but put it together in a novel way that it creates surplus should be analyzed. In small bits and pieces, individuals give up their privacy tranche by tranche. Small amounts of time are spent time on time. People, especially young people, may have a miscalibration about the value of information released about them. Based on hyperbolic discounting myopia, they may underestimate the total future consequences of their share of privacy.

The time spent on social media should become closer subject to scrutiny and the impact on opportunity costs onto the labor market. For instance, countries that ban social media or restrict the Internet, like slowing it down or censoring certain media, could become valuable sources of variance to compare to. Network theories for e-blasting information should become another area of interest to be studied in relation to hyper-hyperbolic discounting fallibilities. Emotional reactions and emotional externalities of communication could be another area of behavioral economics research in the privacy and information sharing predicament domain. The role of attention should be addressed as another moderator variable that is quite unstudied in the digital media era. Thereby, interesting new questions arise, such as how to measure attention—is it the time allocation or the emotional arousal information bestows individuals with?

The preliminary results may be generalized for other user-generated web contents such as blogs, wikis, discussion forums, posts, chats, tweets, podcasts, pins, digital images, videos, tiktoks, audio files, and advertisements but also search engine data gathered or electronic devices (e.g., wearable technologies, mobile devices, 5G and the Internet of things). Certain features of the nudgital society may also hold for tracking data, including GPS, geolocation data, traffic, and other transport sensor data and CCTV images or even satellite and aerial imagery. All preliminary results should be taken into consideration for future studies in different countries to examine other cultural influences and their effects on social class and heuristics.

Innovative means should be found to restore trust in media information and overcome obstacles such as the availability heuristic leading to disproportionate competitive advantages of media controlling parties. As remedies, consumer education should target educating social media users about their rights and responsibilities on how to guard their own and other's privacy. E-ethicality trainings could target at strengthening the moral impetus of big data and artificial ethicality in the digital age. Moral trade-offs between privacy infringements and security should also become subject to scrutiny.

Paying attention to the most recent trends of the demands for ethics of inclusion, Law and Economics interdisciplinary dialogue building and human-artificial intelligence algorithm compatibility are expected to become key advancements in behavioral economics and finance leadership of the future. Ethics of Inclusion should offer a comparative approach to understand the most contemporary responsibility challenges of our time. The idea of fairness and social justice should be elucidated from an ethics perspective in our post-pandemic era. Ethics of Inclusion will embrace diversified potentials that overall empower society. Inclusion should be free from any form of discrimination—may it be direct and obvious or more discrete. The age of digitalization opens gates for searchplace discrimination, which is hardly captured in legal regulation or workplace anti-discrimination laws. Future advancements in behavioral economics and finance leadership may address this abyss of discriminatory actions online and find ways how to avert the negative implications of searchplace black hat strategies. For instance, improving the algorithm-human-interaction gap could aid in quality control over online contents and content removal. The strategic display of information but also the competitive infiltration of online search results with nonsense or unrelated content in Googlewashing should become subject to scrutiny and academic discourse over upholding ethics and merit-based anti-discrimination. Anti-discrimination measures of the future should start to integrate insights about searchplace strategies. Searchplace providers such as search engines but also social media tools and career platforms should help refine search results and work toward wiser and more harmonious human-algorithmic interactions.

Most recent law and economics developments should become the basis for solving practical ethical dilemmas arising in the fair distribution of health care, financial aid, education, digitalization and carrying the burden of climate stabilization. In the distribution of COVID-19 relief, a system change may be accomplished when being guided by the interdisciplinary insights of "Law and Economics" that in particular can address the disparate impact of the disease in order to derive targeted inequality alleviation strategies. Future economic policy research may be inspired by legal expertise on the measurement of disparate impact, which could open up the black box of the neoclassical aggregate economic functions calculus as a measure of economic growth. The combined expertise of Law and Economics in their unique interaction may shed light on disparate impacts of disease, public health and economic relief measures. This may lead to an interdisciplinary framework for crisis alleviation through redistribution that in combination can set the course for a better future in a more compassionate and inclusive world.

Promoting governance through algorism offers novel contributions to the broader data science and policy discussion. Future studies should also be concerned with data governance and collection as well as data storage and curation in the access and distribution of online databases and data streams of instant communication. The human decision-making and behavior of data sharing in regard to ownership should become subject to scrutiny in psychology. Ownership in the wake of voluntary personal information sharing and data provenance and expiration in the private and public sectors has to be legally justified [5].

In the future, institutional forms and regulatory tools for data governance should be legally clarified. Open, commercial, personal, and proprietary sources of information that get amalgamated for administrative purposes should be studied and their role in shaping the democracy. In the future, we also need a clearer understanding of the human interaction with data and their social networks and clustering for communication results. The guarantee of safety of the information and the guarantee of the replacement or service, should a social media fail its function to uphold privacy law as intended, is another area of blatant future research demand. Novel qualitative and quantitative mixed methods featuring secondary data analysis, web mining, and predictive models should be tested for holding for the outlined features of the new economy alongside advancing randomized controlled trials, sentiment analysis, and smart contract technologies. Ethical considerations of machine learning and biologically-inspired models should be considered in theory and practice. Mobile applications of user communities should be scrutinized.

As for consumer-worker conditions, unionization of the social media workers could help uphold legal rights and ethical imperatives of privacy, security, and personal data protection. Data and algorithms should be studied by legal experts on licensing and ownership in the use of personal and proprietary data. Transparency, accountability, and participation in data processing should also become freed from social discrimination. Fairness-awareness programs in data mining and machine learning coupled with privacy-enhancing technologies should be introduced in security studies of the public sector. Public rights of free speech online in the dialogue based on trust should be emphasized in future educational programs. Policy implications of the presented ideas range from security to human rights and law to civic empowerment. Citizen empowerment should feature community efforts to protect data and information sharing to be free of ethical downfalls. Social media use education should be ingrained in standard curricula and children should be raised with an honest awareness of their act of engagement on social media in the nudgital society of the digital century.

Future research may also delve into moderator variables of the utility derived from information sharing and privacy. For instance, extraversion and introversion could be moderating the overall pleasure retrieved from communication or silence. Future research may also address prescriptive recommendations on how to educate individuals about the risks and dangers of information sharing in the digital age. Attention must also be paid to uphold accuracy in times of fake news and self-created social information, especially in competitive settings. Certain societal segments that are not represented strongly online should somehow be integrated into big data in order to democratize the information, which is considered as big data "norm," or standard by which the social media user is measured on.

At the same time, psychologically-guided studies could unravel a predictive approach and validate the outlined ideas by testing the proposed theoretical assumptions in laboratory and field study settings. In particular, the proposed nomenclature's validity could be studied and the percentage of information sharing types captured in the population. The moderator variable age could be phased in as it appears to be conundrum why younger people, who have more to lose given a longer time ahead to live are in particular prone to use new social media and lavishly share their lives in e-blasts to public.

Regarding direct implications, a tax may be used to offset problems of the costs and risks of social media privacy infringements in the big data era.[1] Drawing from utility usually measured by the willingness to pay different amounts of money for different options, laboratory experiments may operationalize the value of privacy by measuring how much money people would be willing to pay for repurchasing their data or having a social media account that can only be viewed but no personal data can be resold or put in context to others. These attempts could also serve as a guideline for policy regulations and free-market solutions. Social media could offer services of having accounts that are private in that sense that no surplus value can be reaped by reselling information or big data storage and computation can occur. This may serve as an indicator of revealed preferences of social media privacy.

The privacy paradox may be scrutinized in behavioral economics laboratory and field experiments. Potential individual influencing factors such as gender, age, trust, and personality differences may be tested in order to retrieve information on how to educate the social media user and regulate the social media provider. Regarding regulation, splitting social media power cartels may be one solution to decrease the big data social media user disadvantage.

Taxation for information sharing may create another incentive to slow down unreflected information share. The tax revenues could be used to offset some of the societal costs of privacy infringement. In addition, fines for privacy infringement could help to uphold e-ethics in the digital age. From the economics perspective, interesting moderator variables for future studies are the distinction between active and passive communication. Further, model robustness checks could follow and learning effects depicted.

Access to information what happens with data and how big data is used appears crucial for learning people as a well-calibration of their relation to their information. Communication costs and benefits are assumed to not be additive and separable, leaving an interesting field for future studies in this domain. An introduction of the concept of "Conscientious Usership" could be pursued—capturing unpaid online usership of search engines, social media, and online marketplaces that gets harvested for big data. The use of future-oriented discounting in the predicament between utility in information sharing versus dignity in privacy should become emphasized. Special attention should be paid to digital social media platforms' role as information exchange on minor medical conditions during the

[1] https://www.nytimes.com/2017/11/14/business/dealbook/taxing-companies-for-using-our-personal-data.html?rref=collection%2Fsectioncollection%2Fbusiness&action=click&contentCollection=business&region=stream&module=stream_unit&version=latest&contentPlacement=8&pgtype=sectionfront.

COVID-19 pandemic—foremost in social media COVID Long Haul groups that currently revolutionize the medical information market but come with the caveat of privacy concerns and discrimination potential of vulnerable patient groups. Data ownership but also the culture of e-ethics that we all create online have to be emphasized in the digital age.

The communication patterns could be classified into different types of communication in the future, e.g., certain node specificities detected, such as communication within a family, with friends, and in hierarchical situations like at work. The absolute and relative influence of information sharers could become part of a network description approach as well. Impact factor measurements could be based on status, search engine rank, and connections to capture global influence. Complexity of information would need to be controlled based on information processing times and time allocation preferences to information, hence attention. Communication costs should be separated in economic models in fixed and variable communication costs in the future and a potential separation between fixed communication costs for social media providers and a variable communication costs for social media users be depicted [19].

Overall, the presented piece can also serve as a first step toward advocating for education about information sharing in order to curb harmful information sharing discounting fallibility. From legal and governance perspectives, the outlined ideas may stimulate the e-privacy infringement regulations discourse in the pursuit of the greater goals of democratization of information, equality of communication surplus, and upheld humane dignity and e-ethics in the big data era.

7.1 Behavioral Economics of Tomorrow

Part of the book addressed the connection of nudging and social class structure in international trade in order to derive conclusions about implicit societal impetus of nudging and winking in the twenty-first century. Alongside providing an overview of behavioral sciences with an application in the public domain, this part took a critical approach in the economic analysis of contemporary public governance through nudging and winking enabled through global e-trade and outsourcing through algorithms around the globe. In this light, governing our common welfare through deceptive means and outsourced governance on social media appears critical. In combination with the underlying assumption of the nudgers knowing better what is right, just, and fair within society, the digital age and social media tools hold potential unprecedented ethical challenges.

Drawing from some of the historical foundations of political economy, this part sought to advance the field of behavioral economics through a critical stance on behavioral sciences and new media use for guiding on public concerns in the digital age. In its totality, this chapter offered a critical analysis of behavioral economics with an emphasis on political economy [9, 10]. By revealing the contradictions of the social media age of the nudgital society, light is shed on the implicit class struggle rooted in the nudgital social relations of production.

Pointing out the limitations of behavioral insights to inform about public choices accurately became the basis of the critique of a certain ruling class nudging a wide populace with the help of social media. An analysis of the process of the circulation of information leads to conclusions about the metamorphosis of big data and their circuit. By shedding light on the inherent class division in those who nudge (the nudgers) and those who are being nudged (the nudged), the book proposed strategies to unravel how the use of behavioral economics for the greater societal good in combination with the rise of social media big data creation may hold unknown socio-ethical downfalls. The part thereby took a heterodox economics stance in order to search for interdisciplinary improvement recommendations on how to alleviate public-sector concerns in the digital age more inclusively.

An introduction to behavioral economics nudging was followed by a description of social hierarchy in the nudgital society. The underlying structures that lead to a class division in those who nudge and those who are being nudged are captured for the first time in order to draw conclusions about the hidden downfalls and risks of the nudgital society. Implications of invisible governance through nudging lead to open questions about ethics in the information age and recommendations for societal and democratic improvement in the twenty-first century. The book closes with a preview of potential future directions of the novel insights gained on the nudgital society. Challenging contemporary behavioral insights theory is aimed at moving together toward a more inclusive future wiser, more self-informed, and protected digital society.

The nudgitalist act becomes problematic when being coupled with infiltration with fake news and alternative facts that curb democratic acts, e.g., manipulating voting behavior. Deepfake is another growing concern online faced by consumers of content on the Internet, which can come in very different forms. Growing use of social online media platforms and an increased participation in content creation online as well as an exponentially-rising self-expression online have grown the problem of deepfakes. Deepfakes include disinformation and misinformation including sound and image manipulation or speed of display altered to overemphasized certain features that gets circulated on social media. With the constant merging of consumers and producers online in self-expression as well as the growing praticipatory nature of online information gathering but also in light of the Internet at information source for hiring and personal choice decisions, misinformation or disinformation has—almost unnoticingly—become a tool of manipulation and alternative reality creation. Deepfakes are the strategic use of synthetic media that creates an alternative reality often by replacing or adding information to an already existing media content. While some contexts and motivations to create deepfakes are more obvious, e.g., in gathering attention or entertainment through humorous content merging; information about other strategic use of content merges and blending in mis- and disinformation are rather hardly described. For instance, in the use of deepfake for hoaxes, cyberbullying and cyberfraud, growing evidence is just emerging on the technicalities. Future research may also address legal implications and socio-economic consequences of how misrepresenations can harm individuals' reputation, career, and personal well-being.

As for constructive ways forward, consent rights to online exhibits of information should be strengthened. A first step was done in the Digital Millennium Copyright Act (DMCA). The Digital Millennium Copyright Act is a 1998 United States copyright law that implements two 1996 treaties of the World Intellectual Property Organization.[2] It criminalizes production and dissemination of technology, devices, or services intended to circumvent measures that control access to copyrighted works. The DMCA affirm first amendment rights when reversing censorware products. The DMCA helps the removal of unwanted copyrighted content online worldwide, whereas in Europe the right to delete is more common to request a deletion of unwanted personal content online.

Ethical considerations concern fair Internet use and searchplace discrimination especially in cases where unfavorable information gets hyped or snippets of search engine descriptions get manipulated in an unfavorable manner in a competitive setting. These searchplace discrimination techniques are rather novel and demand for regulatory attention of online social media, and search engine providers. Preliminary evidence reports that consumers feel cheated if being explosed to deepfake [1]. Presence of too much disinformation elicits reason to be skeptical of genuine information [1].

This book built the case for a revolution in "Conscientious Usership"—capturing unpaid online usership of search engines, social media and online marketplaces that gets harvested for big data. The use of future-oriented discounting in the predicament between utility in information sharing versus dignity in privacy was emphasized. Special attention was paid to digital social media platforms' role as information exchange on minor medical conditions during the COVID-19 pandemic—foremost in social media COVID Long Haul groups that currently revolutionize the medical information market. Patients as vulnerable population sharing information online also come with the caveat of privacy concerns and discrimination potential of impaired groups. Future innovations should be targeted at harvesting the benefits of instant communication but being mindful and respectful to those who are pressured in pain or under the influence of medical drugs searching for relief of medical conditions online.

Data ownership mindfullness trainings but also the culture of e-ethics that we all create online have to be emphasized in the digital age. Special attention should be paid to the role social online media has gained for finance and market communication. Emphasis should also be placed on the most novel developments of search engine results and social online media scraping for hiring and career promotion. Awareness of the role of these online information sources and communication channels will build consumer pressure to design fair online competition as well as ethical marketplaces evolving on the Internet.

[2] https://en.wikipedia.org/wiki/Digital_Millennium_Copyright_Act#Edelman_v._N2H2.

7.2 Discounting

Time determines life. While all humans face the same natural constraints of 24-h days, behavioral economics found individuals differing in discounting preference for immediate rewards over delayed gratification [6, 12, 24, 25]. Regarding monetary gains, individuals were also shown to hold mental accounts dependent on a reference point but also in regard to how to allocate money to cause individuals to care about.

Part of the book was dedicated to time management with a particular focus on financial management. A theoretical part described mental accounting in the behavioral economics domain [32]. Mental temporal accounting was presented to (1) occur for time allocations for recreational time spent on one's own, social leisure time, and work time; (2) discounting variance based on economic, social, and environmental contexts as well as age; (3) how individuals can be nudged into different time perceptions given external cues putting them into a specific mindset.

In empirical investigations, individuals were shown to differ in mental temporal accounts. Decision-makers have natural mental accounts for how to spend 24 h a day, 720 h a month, 8760 h a year, or 700,800 h an average life. Empirical evidence was provided that outlined individuals have implicit mental accounts for how much time to spend on their own, how much time to be allocated toward working, and how much time to just enjoy life in leisure. This mental accounting was proven to also depend on reference points of the age and parenthood. Individuals were outlined to be susceptible to external cues that nudge them into certain mental timeframes that determine their mental time allocation preference. As individuals were depicted to differ in time spent on their own and social time but also vary in their choices of time spent working and leisure time, as a next step we would need to revise contract theory's primary focus on incentivizing with monetary gains and call for attention to a natural mental temporal accounting trade-off decision between work and leisure time. By doing so we could show that the classical mental accounting theory [32] is actually connected to mental temporal accounting how to spend time but also dependent on the reference point of age.

The presented results call for an opening of (1) the behavioral economics idea of the classical mental accounting theory [32]; (2) connecting it to mental temporal accounting; as well as (3) neoclassical contract theory for work–leisure time trade-offs alongside providing future research prospects. Elucidating how contexts and experiencing critical life stages influence temporal activity allocation choices promises to improve manifold choices on education, health, asset management, career paths, and common goods preservation throughout life [2–4, 8, 11, 14, 18, 28, 33–36].

7.3 On the Collective Soul of Economics

In society, language is used to embody theories of reality. Ideas relevant to events such as employment and bargaining, material possession, and physical environment are covered by communication. Linguists see the discourse inseparable from social and economic factors [7]. Different social strata and groups but also different institutions and media have different varieties of language available to them. Linguistic variations reflect and actively express the structured social differences which give rise to inequality and economic dispersion. Beyond effect and reflex of social organization and process, language is part of the social process. Orwell [18] was the first to recognize connections between language, ideas, and social structure in 1984.

Part of the book unveiled linguistic practices which implicitly become instruments to assert social inequality in economic terms. The author thereby studied how linguistic structures are used to explore, systematize, transform, and often create realities about prices and markets. Language was thereby portrayed as a means to regulate the ideas and behaviors of social masses. Access to information was shown to determine the position of the individual in economic market competition. Language was portrayed as implicit determinant to classify and rank people to assert institutional or personal status in society.

Parts of the book sought to shed light on unknown dangers of mass communication of economic correlates that implicitly aid in building expectations and hence bubbles to add to economic fluctuations in price. Media was portrayed as powerful mode of language and thought. Mass communication was viewed as a tool to unconsciously drive the economic engine. While this part of the book departs from standard linguistic thoughts on means to deceive the populace and process of reality control, unprecedentedly-described implicit economic fluctuations built by media representations were the focus of attention. The media was argued as to control society's relationship to material reality [7].

What is the driver of equilibrium and what communication causes fluctuations in the economy? The book showed how communication ends with not conveying the meaning that was wished to communicate but creates a stylized fantasy that is far from pure. The communication was shown to be repackaged at the recipient, whose experience with others influences the way information is processed and guides actions. In this light freedom of speech and access to information appear as for holding disastrous outcomes.

For linguistics, the book addresses the meaning and implicit meta-meanings of words and the economic correlates which are built upon messages. Unveiling the economic machinery of language as driver of economic ups and downs thereby aids in deriving communication recommendations to ease economic fluctuations. Journalists can be enabled to understand the economic ethos of words and the moral imperative of their economic coverage.

While the economic market actor has freedom of choice, the work highlights the interdependence of market actors in prices, which are determined by communication creating expectations or contracting collective choice. The use of language as control was shown to be limited by the collective reality of price and the irrationality of the crowd. In this sense, language determines the social structure, inequality, and the distribution of power within society.

Society was outlined as organized upon a principle of unequal power through the form of public communication. Newspapers, governments, bureaucracies, and intellectuals control through language. Using linguistic analysis as a way of uncovering the making of economic booms and busts affects the general consciousness about language as an implicit economic correlate and basic of economic fluctuations. Contemporary writings address political, institutional, and social processes as makers of crises, yet the role of information for the building of economic moods is completely neglected.

Speech and writing express the social circumstances in which language occurs. While linguists attribute language as a product of social circumstances, the book also shed light on the relationship between social processes forming language. Language shapes social processes. As part of social process, language influences groups. The information available to certain groups determines their states.

While linguists primarily describe the use of language to control the behavior of groups, this part focused on the unintended and unforeseeable consequences of information on price and economic correlates. While there are writings on the role of information to control behavior, less is known about the unintended consequences of information in expectation building and hence bubble creation. Language reflects inequality.

All language is addressed to someone; it is relations and asymmetrical in that sense there are differences in class or states. The relationship is competitive negotiation for power. As language supplies the models and categories of thought, it shapes people's experience of the world. Economic uncertainty may stem from fluctuations of discourse in the media creating a perceived uncertainty. The linguistic equivocation mirrors the tension of the real situation in which people find themselves. Results were presented as a beacon to gain certainty about the economic situation based on the linguistic analysis results of market price creation.

This book took a different interpretation of classic economics treating consumers as passive in terms of prices that are given, who try to maximize their profits and limit costs given a budget constraint. The firm-based theory that takes prices as given searching to maximize profits subject to technological constraints was abandoned for departing from the orthodoxy.

The book acknowledged exchange takes place in disorder, in which prices are the monetary expression of a commodity's quantitative worth. Profit was an objective measure subject to constant scrutiny by the firm's managers, the stock market, the banks, and the public in general. According to Kalecki's [13] Theory of Price, the price of individual firms depends on the relative size of firms due to union power of employees, industry prices, and average costs and monopoly power that determine the markup structure. The uniform price is assumed to be supremely responsive to the market demand and supply.

Actual decisions are always in terms of current and expected market prices. Market prices gravitate in a turbulent manner around prices of production. In keeping with the price setting and cost-cutting behavior of real competition, firms are forced to select the lowest cost reproducible conditions for production. Given real markets are always turbulent, all choices must be "robust" in the sense that they remain valid in the face of normal fluctuations in costs, prices, and profitability.

The profit motive is the dominant factor in the regulation of economic growth [29]. Expected profitability at the core of any economic activity implies a mode of interaction between aggregate demand and supply. Profit thus regulates both—supply and demand [29]. Profit therefore determines an endogenous "natural" rate of growth. The interest rate is the price of finance, financial firms exist to make profit, and competition makes the profit rate of the regulating financial capitals gravitate around the general rate of profit [23].

Profit drives capitalism. If profit fails, the economy is in shock and its capital gains to atrophy. This book also reflected on the gained communication strategy results based on economic classics such as Marx's theory of capitalism, Keynes' [15] liquidity theory underlying aggregate consumption, Sraffa's prices of commodities [31], the Austrian economic school of business cycle theories [26, 27] and Soros' Theory of Reflexivity [30].

7.4 Public-Sector Implications

The wider impetus of this book is to build a scientific foundation for the politics of modern social media use and reveal the law of motion of the modern nudgital society. The nudgital hierarchy unfolds based on the social media mode of production. The use value is thereby the utility derived from sharing and receiving information. The use value is thereby the utility of consuming social media, the want-satisfying power of a good in the classical political economy sense. The exchange value the social media industrialist-capitalist gains from access to information that can be computed into big data that allows deriving behavioral insights for markets and governance technocrats. The information released by the consuming workers becomes an exchange value in big data sold to marketers and technocrats to gain information about nudgebility of potential consumers and general populace. The exchange value is retrieved from simple information circulation and the circulation of value as nudgital, the power to nudge.

The machinery of the social media industrialist-capitalist is the social media users' production as self-acting automata and the consumer-worker gaining gratification through likes and delving into a fantasy world of the self. Information amalgamation can become an act of critical social consequence, which is capital-oriented toward making use of use value in the form of exchange value in the materialization of information. The nudgitalist exploitation begins when information gets turned against individuals. A Gestalt through the bits and pieces of the individually-shared information creates an unfavorable condition in terms of consumer exploitation and in front of governmental authorities.

In all of this, people end up in a subordinate position of their information being turned against them within the digital society. Not only does the social media consumer-workers' dependence on communication media infringes on their privacy. Their social relations get exposed, and they lose their privacy to those owners of the means of social media production. Information can also be used to nudge people into making choices, may that be consumer decision, or may that be their political choices.

The nudgitalist exploitation also holds when technocrats use heuristics and nudges to create selfish outcomes or undermine democracy. Ethical abysses of the nudgital society open when the social media is used for public opinion building and public discourse restructuring. Social media not only allows one to estimate target audience's preferences and societal trends but also imposes direct and indirect influences onto society by shaping the public opinion with real and alternative facts. Government officials' gaining information about the populace that can be used to interfere in the democratic voting process, for instance, in regard to curbing voting behavior or misinformation leading people astray from their own will and wishes. The social intertwining of the media platform and the democratic act of voting has been outlined in political votes that were accused to have been compromised by availability heuristic biases and fake news. Data can also be turned against the social media consumer-worker by governance technocrats for the sake of security and protection purposes, for instance, social media information can be linked together for tax verification purposes.

Governments have been transformed under the impact of the digital revolution. Instant information flow, computational power and visualization techniques, sophisticated computer technologies, and unprecedented analytical tools allow policy-makers to interact with citizens more efficiently and make well-informed decisions based on personal data. New media technologies equip individuals with constant information flows about informal networks and personal data. Novel outreach channels have created innovative ways to participate in public decision-making processes with a partially unknown societal impact at a larger scale, scope, and faster pace than ever before. Big data analytics and the Internet of things automate many public outreach activities and services in the twenty-first century. Not only do we benefit from the greatly increasing efficiency of information transfer, but there may also be potential costs and risks of ubiquitous surveillance and implicit persuasion means that may threaten democracy.

The digital era governance and democracy features data-driven security in central and local governments through algorithmic surveillance. Open-source data movements can become a governance regulation tool. In the sharing economy, public opinion and participation in the democratic process have become dependent on data literacy. Research on the nudgital society holds key necessary information about capacity-building and knowledge sharing within government with respect to certain inalienable rights of privacy protection. The nudgital society's paradox that information sharing in the social compound gets pitted against privacy-protecting alienation requires an ideological superstructure to sustain and tolerate hidden exploitation.

All these features are modern times phenomena as the technology and big data creating computational power are currently emerging. The transferability of the commodity of information itself, hence the big data amalgamation over time and space to store, package, preserve, and transport information from one owner to another appears critical. The legal leeway to allow private information sharing implicitly leads to individuals losing their private ownership rights to the commodity of information upon release on social media and the right to trade information. The transferability of these private rights from one owner to another may infringe on privacy protection, human rights, and humane dignity upholding mandates.

Not only pointing at the ethical downfalls of the nudgital society, but also defining social media users as workers is of monumental significance to understand the construction of the nudgital society and bestow upon us social media consumer-workers labor rights. The technical relationship between the different economic actors is completely voluntary and based on trust [20–22]. The creation of use value is outsourced to the community (e.g., in likes), and the share of information about the workers from the social media capitalist to the market or nudgitalists remains without a clear work contract and without protection of a labor union. The worker–employer relationship needs to be protected, and a minimum wage should settle for the market value of the worker producing during the working day. Wages would be needed to maintain their labor power of the workers minus the costs of the production. Unpaid laborers should not only be compensated for their opportunity costs of time but should enjoy the workers' privilege of right to privacy, prevention of misuse of the information they share, and have the right to access to accurate information but also protection from nudging in the establishment of the right to voluntarily fail.

The nature of making profit from information in exchange value is questionable. Information exchange of the industrialist-capitalist is different than neoclassical goods and services trade insofar since for the capitalist-industrialist making money off privacy and the consumer-workers share of information without knowledge and/or control over the recipient of the amalgamated mass of privacy released. Workers are never indifferent to their use value and their inputs may also produce unfavorable outcomes for them. The exchange value will sell for an adequate profit and is legally permitted, yet it can destroy the reputation and standing as well as potentially the access of the individual to country entrance if considering the proposed social media information release mandate at border controls. Care must be taken for privacy infringement and the product of amalgamated big data and how useful it is for society.

7.5 Legal and Global Governance Implications

By shedding light on these risks of the social media age and the implicit dynamism of capitalism forming around information, a social formation of social media workers' right can be heralded. Social media user-workers should be defined to hold inalienable rights to privacy and be forgotten, to be protected from data misuse of information share, they should be granted the right to access accurate information and—in light of the nudgitalist audacity—the right to choose and to fail.

People's right to privacy and to be forgotten: The transformation of a use value into a social use value and a commodity has technical, social, and political preconditions. Information gets traded and ownership of privacy is transferred in information sharing. Upon sharing information on social media, the consumer-worker bestows the social media capitalist-industrialist with access to previously-private information. The social media capitalist then transforms the information into useful value by offering and selling the bundled information to nudgitalists, who then can draw inferences about certain consumer group preferences and guide their choices.

Overall, the nudgital society leads to a dangerous infringement upon the independence of individuals in their freedom of choice and a social stratification into those who have access to the amalgamated information of social media consumer-workers. There is a trade-off between communication and privacy in an implicit contract of the use of personal data. Power is exercised through the accumulation of information, including the quality of insatiability of social media consumer-workers to constantly upload information and the social media capitalist-industrialist reaping profits from selling it. No international publication exists to this date about the most novel facets of digital inequality around the world. With the most recent developments during COVID-19 of patients searching information online for relief, especially those who suffer from chronic long-term conditions, rights to be forgotten online need to be further developed in the practical implementation to protect these vulnerable populations.

Social media thereby reveals to hold a sticky memory that allows storage of information in the international arena eternally. Privacy and information share regulations depend on national governments. For instance, in the commodification of privacy, the EU is much more cautious with consumer privacy than that of the US. Data protection and commercial privacy are considered fundamental human rights to be safeguarded in Europe and a basis of dignity. In contrast, the US approach toward commercial privacy focuses on only protecting the economic interests of consumers. Current privacy regulations are considered as not sufficient in targeting actions that cause non-economic and other kinds of harm to consumers. While in Europe healthcare data is public, in Canada, there is a public interest to make the data more public. Especially when it comes to health data, the coronavirus crisis has transformed privacy around the world, with the vulnerability of patients being a priority to be advocated for.

However, privacy and information sharing guidelines appear to be culturally-dependent phenomena. Information about privacy boundary conditions can be obtained from the transatlantic dialogue between the US and Europe on privacy protection. The EU's privacy approach is based on Articles 7 and 8 of the Charter of the Fundamental Rights of the EU, which grants individuals rights to protection, access, and request of data concerning him- or herself. European privacy is oriented around consumer consent. The 2016 EU General Data Protection Regulation (GDPR) ruled the right to be forgotten under certain circumstances in the online world.

Consumer consent and dealing with incomplete, outdated, and irrelevant information is legally regulated. GDPR establishes regulatory fines for non-complying companies applicable to foreign companies whose data processing actions are related to "good and services" that they provide to data subjects in the EU, so also including US companies operating in the virtual space accessible by European citizens. The EU privacy approach offers member states flexibility in data management for national security and other exceptional circumstances but also protects civilians from common potential circumstances for data abuse, while there are standardized data management policy procedures regardless of a companies' country-of-origin or operational locations. The EU's privacy approach has higher regulatory costs, is not specified by sectors, and the right to be forgotten still needs enforcement validity.

The US approach to privacy is sector-specific. Commercial privacy is pitted against economic interests and neither seen as civil liberty nor as constitutional right. US privacy is regulated by the Federal Communications Commission (FCC) and the Federal Trade Commission (FTC). Overall in the US, the general definitions of unfair and deceptive practices give the FTC a wider scope for monitoring and restricting corporate privacy infringements. The FTC has a wide variety of tools for data protection, yet the responsibility is split between the FTC and the FCC, which increases bureaucratic and regulatory costs and limits industry oversight.

While the EU framework treats commercial privacy as a basic human right leading to a more extensive protection of individual's privacy including data collection, use, and share, the EU framework is also non-sectoral and allows sovereign nation states to overrule common data management policies for the sake of national security and protection. The US framework lacks a centralized privacy regulation approach, yet is sector-specific but split regarding oversight in the domains of the FCC and FTC.

People's right to prevent misuse of information they share: By US standards, social media is required by the FTC to ask users for permission if it wants to alter its privacy practices. Under Sect. 5 of the FTC Act that states that (1) unfair practices are caused or are likely to cause substantial injury to consumers or cannot reasonably be avoided by consumers; and (2) deceptive practices are practices that likely are misleading or actually misleading the consumer.

The August 2016 court rule decision of WhatsApp being enabled to share more user data, especially user phone numbers, with Facebook in order to track customer-workers' use metrics and targeted user advertising faced a huge backlash in the EU, where data sharing was ordered to be halted and Germany deemed its practices as illegal. In the US, the Federal Trade Commission (FTC) began reviewing joint complaints from consumer privacy groups. The recent WhatsApp data sharing is a possible violation of this requirement since it only allowed consumers to opt-out of most of the data sharing while lacking clarity and specificity. WhatsApp's restrictive opt-out option and incomplete data sharing restrictions were argued to be perceived as unfair and deceptive (Tse, in speech, March 25).

People's right to access to accurate information: In the nudgital society, profits appear in the circuit of information and take on different forms in the new media age. The possibility of trading information and reaping benefits from information sharing of others determines the unequal position of people in society. The possession of knowledge stems from the surplus derived from the activity of production, hence the information share of social media consumer-producers. This confrontation of labor and consumption is not apparent in the modern marketplace. The class division remains quite invisible in the implicit workings of the system.

Ethical questions arise if there is a transparency about the capitalist's share of information and a fair social value benefits distribution among the capitalist and the worker. In addition, under the cloak of security and protection, privacy infringements by sharing information with the nudgitalist are questionable. By outlining these market procedures, fairness in the distribution of gains should be accomplished and privacy infringing information sharing limited or at least guided by the legal oversight. Access to information about the storage, preservation, packaging, and transportation of data is non-existent, demanding more information about behind-the-scenes' social media conduct. Transforming private information from use value to exchange value is an undisclosed and therefore potentially problem-fraught process that holds implicit inequality within itself.

From a societal standpoint, also the missing wealth production in the social media economy appears striking. Thereby, the dangers of information release and transfer and the hidden exchange value accrued on the side of the media innovator are left unspoken. The importance of shedding light on such, though, is blatant as for stripping the populace from inalienable rights of privacy while reaping benefits at the expense of their susceptibility. Nudges in combination with misinformation and power abuse in the shadow of subliminal manipulation can strip the populace from democratic rights to choose and voluntary fail.

People's right to choose and fail: In the personal information sharing age and nudgital society, attention must be given to privacy and human dignity. The nudgital society opens a gate to gain information about consumer choices and voting preferences. The uneven distribution of key information about people's choices opens a gate to tricking people into choices. The so-called nudging attempt though raises ethical questions about human dignity and the audacity of some to know better what is better for society as a whole. Because governance is a historical process, no one person can control or direct it, thereby creating a global complex of governance connections that precedes the individual administration.

Structural contradictions describe the class struggle between the nudged in opposition to the nudgers in the nudgital society. Since societal actors who involuntarily are nudged are separated from an active reflection process when being nudged, the moral weight is placed on the nudger. Though democratically-elected and put into charge, the nudgers' checks-and-balances of power seem concentrated and under disguise through the middleman of social media capitalist-industrialists who collect information. Rather than focusing on how to trick people into involuntary choices, the revelations should guide us to demand to educate people on a broad scale about their fallibility in choice behavior.

In a self-enlightened society, people have a right to voluntarily fail. Nudging implies a loss of degrees of freedom and disrespect of human dignity; hence, the nudgital society will lead to structural contradictions. Their rational and efficient thinking and voluntary engagement in governmentally-enforced action become divorced from rational and efficient reflection. No one entity should decide to control or direct other's choices, thereby creating a global complex of social connections among the governed for the sake of efficiency for the common good. The economic formation of human decision-making in society should never precede the human voluntary decision.

People's right to not be forgotten: Besides the right to be forgotten in the European context and the Digital Millennium Copyright Act in the US, a right to not be forgotten should be advocated for in case online content about oneself gets removed online in a competitive act. Rights to not be forgotten come in automatic algorithmic responses to potentially-false competitive claims to remove content from the Internet. For instance, competitors may try to push down competing products or information about competitors displayed online. In searchplaces like Google, Bing, and Yahoo! a prominent place in search results has leveraged into a marketable lucrative asset.

A prominent searchplace slot has become a vital market revenue source as it attracts customers for online sales. Search engine optimizers can be purchased that move content up the search engine result hierarchy. Financing online webhosts also helps as do paid Google corporate accounts. When it comes to online competition for prominent searchplace results, so-called white hat strategies are used to push one's content up in searchplaces. So-called blackhat strategies are legal but ethically questionable strategies to unfavorably push down others' content in competitive settings. For instance, pegging information to other big data clouds can be one strategy to make content disappear when searching for it. Another way to remove content online is to stag negative content to the big data cloud of someone or flag content as inappropriate, even if they are appropriate. Since most quality control is outsourced to algorithms in searchplaces, once flagged, a content gets removed automatically and it may be forever removed or it takes time to reappear in a non-prominent position in searchplaces. Here a right to not be forgotten should be advocated for in order to have a better quality control what comes online and what stays online even if being flagged as inappropriate inappropriately.

The use of clickfarms is another ethically-challenging way to peg negative contents into big data clouds in order to make someone's content being pushed down in searchplaces. For instance, outdated information but also information of unsolved crimes is unfavorable to have in one's big data, which gets then downgraded to be displayed by searchplace algorithms. Information on less credible pages displayed also wears down the credibility of one's own online content. Sending in removal request of personal information from less credible pages and privacy infringement announcements advocating for the removal of information from less credible pages is key in cleaning up one's big data and maintaining an accurate and healthy online appearance.

Clickfarms and algorithms overrating specific pages are also used in a competitive context to overplay certain information online. For instance, if someone has a review that is unfavorable, blackhat strategies can manipulate this one review to stick out. It is thus advised to advocate for consumers being given more autonomy and power to discuss and craft their Internet appearance with searchplace providers. To this day, most search engine providers hardly offer any opportunity to communicate with algorithm engineers and they hardly provide information on how to correct algorithmic errors that can harm the individual's reputation, career, and public perception and reputation. Consumer protection agencies are advised to be alerted for the trend of searchplace discrimination and search engine failures echoing in societal downfalls to increase with the rising levels of Internet consumption for professional purposes and meta-world creation.

Overall, there is an inherent inequality of social positions, manifested primarily in the respective capacities of reaping benefit from amalgamated information and being a vulnerable consumer on the other hand, which leads to a disparity of social positions online. The distribution of power leads to a natural order of human activity, in which the nudgers are in charge of nudging the populace. Moral value is separated from economic value, and hence placing the fate of the populace into the arms of the behavioral economists raises problems of lack of oversight and concentration of objective economic value rule in the nudgital society.

With the communication on the nudgital society just having started, it remains on us to redesign the apparatus of production in ways that make impossible the infringement on private information through the natural tendency to share information, care about others, and express oneself. Governance crises are rooted in the contradictory character of the value creation through big data. The formation of value is a complex determination and we still need more research to understand the deep structures of market behavior in the digital age.

References

1. Adkar, C. (forthcoming). Recognizing inauthenticity on the internet: How to deal with deepfakes? In *Proceedings of the 27th RAIS Research Association for Interdisciplinary Sciences Conference on Social Sciences and Humanities* (June 2022).
2. Arrow, K. (1969). Classificatory notes on the production and transmission of technological knowledge. *American Economic Review, 59*(2), 29–35.
3. Ashraf, N., Karlan, D. S., & Yin, W. (2006). Tying Odysseus to the mast: Evidence from a commitment savings product in the Philippines. *Quarterly Journal of Economics, 121*(2), 635–672. https://doi.org/10.1162/qjec.2006.121.2.635
4. Beshears, J. L., Choi, J. J., Laibson, D. I., Madrian, B. C., & Sakong, J. (2011). *Self-control and liquidity: How to design a commitment contract* (RAND Working Paper Series, WR-895-SSA). Retrieved at SSRN: http://ssrn.com/abstract=1970039 or https://doi.org/10.2139/ssrn.1970039

5. Donahue, J. D., & Zeckhauser, R. J. (2011). *Collaborative governance: Private sector roles for public goals in turbulent times.* Princeton University Press. https://doi.org/10.1515/9781400838103
6. Estle, S. J., Green, L., Myerson, J., & Holt, D. D. (2007). Discounting of money and directly consumable rewards. *Psychological Science, 18,* 58–63. https://doi.org/10.1111/j.1467-9280.2007.01849.x
7. Fowler, R., Hodge, B., Kress, G., & Trew, T. (1979). *Language and control.* Routledge.
8. Gine, X., Karlan, D., & Zinman, J. (2009). *Put your money where your butt is: A commitment contract for smoking cessation* (Policy Research Working Paper Series 4985). The World Bank.
9. Heilbroner, R. (1988). *Behind the veil of economics: Essays in the worldly philosophy.* Norton.
10. Heilbroner, R. (1999). *The worldly philosopher.* Simon & Schuster.
11. Houser, D., Schunk, D., Winter, J., & Xiao, E. (2010). *Temptation and commitment in the laboratory* (IEW Working Paper iewwp 448). Institute for Empirical Research in Economics, University of Zurich.
12. Kahn, J. (2005). *The economic approach to environmental resources.* Thomson-South-Western.
13. Kalecki, M. (1942). A theory of profits. *The Economic Journal, 52*(206/207), 258–267.
14. Kaur, S., Kremer, M., & Mullainathan, S. (2010). Self-control and the development of work arrangements. *American Economic Review, 100*(2), 624–628. https://doi.org/10.1257/aer.100.2.624
15. Keynes, J. M. (1936/2003). *The general theory of employment, interest and money.* Harvard University Press.
16. Kondratiev, N. (1925/1984). *Long wave cycles.* Dutton.
17. Orwell, G. (1949). *1984.* Harcourt Brace.
18. Ostrom, E. (1990). *Governing the commons: The evolution of institutions for collective action.* Cambridge University Press. https://doi.org/10.1017/CBO9780511807763
19. Prat, A. (2017). *Lecture notes 'industrial organization'.* Columbia University.
20. Puaschunder, J. M. (2015). Trust and reciprocity drive common goods allocation norms. In *Proceedings of the Cambridge Business & Economics Conference.* Cambridge University. *Proceedings of the 2015 6th International Conference of the Association of Global Management Studies at Alfred Lerner Hall of Columbia University.* The Association of Global Management Studies. *Oxford Journal: An International Journal of Business & Economics.*
21. Puaschunder, J. M. (2016a). Global responsible intergenerational leadership: The quest of an integration of intergenerational equity in Corporate Social Responsibility (CSR) models. *Annals in Social Responsibility, 2*(1), 1–12. https://doi.org/10.1108/ASR-10-2015-0016
22. Puaschunder, J. M. (2016b). Intergenerational climate change burden sharing: An economics of climate stability research agenda proposal. *Global Journal of Management and Business Research: Economics and Commerce, 16*(3), 31–38.
23. Puaschunder, J. M. (2019). Long-term investments. In L. W. Filho. A. Azul, L. Brandli, P. Özuyar, & T. Wall (Eds.), *Partnerships for the goals: Encyclopedia of the UN sustainable development goals.* Springer Nature.
24. Rubinstein, M. (2003). Great moments in financial economics: I, present value. *Journal of Investment Management, 3*(4), 87–103.
25. Samuelson, P. (1937). A note on measurement of utility. *The Review of Economic Studies, 4,* 155–161. https://doi.org/10.2307/2967612
26. Schumpeter, J. A. (1934). *The theory of economic development.* Harvard University.
27. Schumpeter, J. A. (1951/1989). *Essays on entrepreneurs, innovations, business cycles, and the evolution of capitalism.* Transaction.
28. Sen, A. K. (2002). Goals, commitment, and identity. In A. Sen (Ed.), *Rationality and freedom* (pp. 15–32). Harvard University Press.
29. Shaikh, A. M. (2016). *Capitalism: Competition, conflict, and crises.* Oxford University Press. https://doi.org/10.1093/acprof:oso/9780199390632.001.0001
30. Soros, G. (2003). *The alchemy of finance.* Wiley Finance.
31. Sraffa, P. (1960). *Production of commodities by means of commodities.* Cambridge University Press.

32. Thaler, R. H. (1999). Mental accounting matters. *Decision Making, 12,* 183–206. https://doi.org/10.1002/(SICI)1099-0771(199909)12:3<183::AID-BDM318>3.0.CO;2-F
33. Thaler, R. H., & Sunstein, C. (2008). *Nudge: Improving decisions about health, wealth, and happiness.* Yale University Press.
34. Trope, Y., & Fishbach, A. (2000). Counteractive self-control in overcoming temptation. *Journal of Personality and Social Psychology, 79*(4), 493–506. https://doi.org/10.1037/0022-3514.79.4.493
35. Trope, Y., & Fishbach, A. (2004). Going beyond the motivation given: Self-control and situational control over behavior. In R. Hassin, J. S. Uleman, & J. W. Bargh (Eds.), *The new unconscious* (pp. 537–565). Oxford University Press.
36. Tversky, A., & Shafir, E. (1992). Choice under conflict: The dynamics of deferred decision. *Psychological Science, 3*(6), 358–361. https://doi.org/10.1111/j.1467-9280.1992.tb00047.x

Conclusion 8

The book presents a novel idea in connecting behavioral economics with political economy. Behavioral insights are critiqued insofar as the contemporary behavioral economics approach and extension of behavioral economics application to public-sector problems are shown to hold an implicit underlying social stratification. Democratically-empowered nudgers decide without democratic consent whom to nudge, how to nudge, and what for to nudge. The potential impact of the book for public concerns is to raise a democratic feasibility check of nudging being in line with constitutional values and informed consent demands. In this challenge of the mainstream behavioral economics idea also lies a direct contribution to scientific knowledge spearheading social dominance theory.

The marketability of results demands oversight of the nudgers and well-informed decision-making in an inclusive society. Future research directions and policy recommendations are given that aim at informing scientific audiences, helping public policy specialists, and empowering the general populace about behavioral economics and their rights to privacy. Challenging concerns about libertarian paternalism lead to the quest for people having the right to fail.

In sum, the book should serve as preliminary first introduction of a behavioral economics critique and application of political economy and social dominance theory in the behavioral insights domain for the sake of protection of people's rights to privacy. The aim is to provide insightful information on how society can make rational and efficient decisions in order to maximize welfare without losses of privacy and unjust reaping of undisclosed workers.

In the digital age, capitalism has been built into free social media cyberculture. The forces of production on social media create a sociotechnical apparatus open to change under the dynamics of capitalism. Yet with economics being primarily focused on prices in markets, the social process of exchange of information has been left with little attention as being perceived as a naturally given fact. Future studies should draw attention to these implicit dependency and how commercial relations shape the sphere of personal time use and consumption as well as societal

J. Puaschunder, *Advances in Behavioral Economics and Finance Leadership*,
Contributions to Economics, https://doi.org/10.1007/978-3-031-15710-3_8

standing. The laws of motion of information are believed to differ from the laws of requirements of capital accumulation, yet have to be studied further to be fully explained. To date, no economic model exists that captures the implicit utility of voluntary information share for the sake of engaging on social media. The use value in such differs from the standard neoclassical notion of utility as not subjectively determined by the buyer of a good. In addition, the consumer of social media is not aware of having become a worker in an implicit contract of joy derived from the interaction on the social media tool and sacrificing his or her productive labor time on the social media virtual market space. The changed social media hierarchy demands a resolution in studying these entirely new modes of production. Social media identities differing from our real identities or anonymous social media use are proposed alternatives. Given this novel insight, we need to start estimating the value utility function of the nudged to release information to nudgers and relate it to the economic surplus the nudgers can reap from putting information together and market it to governance professionals.

Artificial intelligence poses historically unique challenges for organizational settings. In a world, where there is a currently ongoing entrance of AI and algorithms into the workforce, the emerging autonomy and superiority of AI holds unique potentials but also economic and ethical challenges in the organizational context. With AI being endowed with quasi-human rights and citizenship in the Western and Arabic worlds, the question arises, how to handle overpopulation but also misbehavior of AI? Should AI become eternal or is there a virtue in switching off AI at a certain point? If so, we may have to redefine laws around killing, define a virtue of killing, and draw on philosophy to answer the question of how to handle the abyss of killing AI with ethical grace, rational efficiency, and fair style. Further, market disruptions in the wake of AI may already have begun, pressing for a demand to monitor and alleviate potential downfalls of AI. Lastly, eternal AI overpopulation problems and resource consumption demand for attention on the impact of AI on sustainable development. The presented theoretical results will set the ground for a controlled AI evolution in the twenty-first century and guide on the entrance of AI into our contemporary workforce in the organizational world.

As emerging globally trend, AI is extending its presence at almost all levels of social conduct and thereby raised high expectations but also grave concerns [10, 34, 39]. With the dramatic growth in diversity and entrance of emerging technologies in today's societies, such as algorithms, social robots, lifelike computer graphics (avatars), and virtual reality tools and haptic systems, the social complexity of these challenges is on the rise [24]. One of the main challenges in developing and applying modern technologies in our societies is the identification and consideration of ethical issues surrounding AI [24]. The call for AI ethics (AIE) has emerged [25–31]. A growing number of AI and robotics researchers have demanded to create a framework on AI ethics building on the benefits of humanities, philosophy, natural sciences, sociology, and social neuroscience.

AI will hold the potential to replicate human existence but also grant eternal being opportunities. In the eye of overpopulation concerns, finding mechanisms to switch off AI would be a solution to avoid a crowding of the planet. But AI currently also reaches quasi-human status through actual personhood—e.g., via citizenship and quasi-human rights applied in the Common Law but also Roman Law territories of the US and the EU. Leveraging AI entities to the status of being through the attribution of legal personhood raises challenging legal and ethical questions. Programming AI to switch itself off or instigating human to switch off AI at a certain point to curb overpopulation but also as quality control against harmful behavior arising out of AI, thereby appears critical as it would come close to suicide or killing. The currently-ongoing market disruption through AI encroaching our workforce raises important organizational behavior concerns. A novel predicament between eternity and overpopulation also calls for thinking about the impact of robots and algorithms on our common sustainability approach.

When considering humans' opportunity to determine life and death of AI, humankind will see the opportunity of AI evolution understood as a human-made evolution determining what contents survive and what to die following the goal to improve the overall offspring and general well-being of humankind. Future managers and corporate leaders will have to determine how to blend in the use of algorithms, robots, and AI on a large scale. Thereby, natural behavioral laws of ethics may serve as a first anchor to determine respectful yet useful conduct around AI.

Lastly, the artificial age imposes challenges on our resource consumption and overpopulation endeavors with regard to the sustainable development goals. The proposed frame will offer innovative insights for corporate conduct regarding artificiality.

This book provides a first guide for organizational behavior specialists to welcome the introduction of AI into our contemporary workforce. The novel and multidisciplinary area of socio-cognitive robotics, and the ethical challenges of emerging technologies are explored. Key ethical features based on past and present research in a variety of AI areas were presented.

As the most novel trend, AI, robots, and algorithms are believed to soon disrupt the economy and employment patterns. With the advancement of technologies, employment patterns will shift to a polarization between AI's rationality and humanness. Robots and social machines have already replaced people in a variety of jobs—e.g., airports smart flight check-in kiosks or self-check-outs instead of traditional cashiers. Almost all traditional professionals are prospected to be infused with or influenced by AI, algorithms, and robotics.

Robots have already begun to serve in the medical and healthcare profession, law, and—of course—IT, transportation, retail, logistics, and finance, to name a few. Social robotics may also serve as quasi-servants that overwhelmingly impact our relationships and workforce. Already, social robots are beginning to take care of our elderly and children, and some studies are currently underway on the effects of such artificial care [1].

Not only will AI and robots offer luxuries of affordability and democratization of access to services as they will be—on the long run—commercially more affordable and readily available to serve all humanity, but also does the longevity potential of machines outperform any human ever having lived [17, 18]. However, the new technology also comes with the price of overpopulation problems and the potential for misuse and violent action. Just like many other technologies, robots could be misused for wars, terrorism, violence, and oppression [1].

AI's entrance in society will revolutionize the interaction between humans and AI with amply legal, moral, and social implications in the organizational context [20, 21]. Autonomous AI entities are currently on the way to become legal quasi-human beings, hence self-rule autonomous entities [37]. AI is in principle distinguished between weak AI, where "the computer is merely an instrument for investigating cognitive processes" and strong AI, where "[t]he processes in the computer are intellectual, self-learning processes" [41, p. 10]. Weak AI is labeled as artificial narrow intelligence (ANI), while strong AI is further distinguished between artificial general intelligence (AGI) and artificial super intelligence (ASI).

The emergence of robotics technology is developing much quicker than previously thought. Robots are anticipated to soon be as ubiquitous as computers are today [24]. Society has long been concerned with the impact of robotics technology from nearly a century ago, when the word "robot" was devised for the first time [9, 24]. The EU Committee on Legal Affairs [15, p. 4] holds that "[U]ltimately there is a possibility that within the space of a few decades AI could surpass human intellectual capacity in a manner which, if not prepared for, could pose a challenge to humanity's capacity to control its own creation and, consequently, perhaps also to its capacity to be in charge of its own destiny and to ensure the survival of the species." AI mimicking human intellect could soon surpass humans intellectually but also holistically breaking the barrier of human-controlled automization, which will become a difficulty in the organizational behavior context [33, 37].

A predicament will soon exist between treating robots in a humane way to uphold social norms within the organizational context as well as reaping benefits from robots [31]. While these machines' value should be extracted, it can be seen critical to anthropomorphizing AI as for ethical concerns to uphold dignity and social norms within the organizational context. The European Union has therefore published a guideline called BS 8611:2016 Robots and robotic devices: Guide to the ethical design and application of robots and robotic systems to protect and anchor of behavioral conduct in the organizational setting. An additional study of the European Parliament investigated the impact of algorithms in the healthcare sector with particular attention to privacy concerns and transparency in the big data age.

Modern literature about robots features cautionary accounts about insufficient programming, evolving behavior, errors, and other issues that make robots unpredictable and potentially risky or dangerous [3–5, 24]. "Observe, orient, decide, act" will therefore become essential in the eye of machine learning autonomy and AI forming a new domain of intellectual entities [2, p. 52, 12, 16, 23]. The uncertainty surrounding AI development and self-learning capabilities give rise to the need for guarding AI and an extension of the current legal system and organizational guidelines to cope with AI [37].

The human perception of and interaction with robot machines with a higher quality physical appearance differs from interaction with a computer, cell phone, or other smart devices [24]. In the organizational context, already now Google glasses and lenses are used in order to collect information from workers. For robotics technology to be successful in a human-driven environment, robots do not only need to meet a level of strength, robustness, physical skills, and improved cognitive ability based on intelligence but should also fulfill a social impetus and ethical conscientiousness. The design and construction of social robots face many challenges; one of the most important is to build robots that can comply with the needs and expectations of the human mind with cognitive capabilities coupled with social warmth [24].

While we have social-cognitive robotics (SCR) as a transdisciplinary area of research and basis for the human-centered design of technology-oriented systems to improve human knowledge functions, judgments and decision-making, collaborations, and learning; hardly any information exists on socio-evolutionary comparisons [24]. Social-cognitive robotics has been evolving and verified through a series of projects to develop advanced and modern technology-based systems to support learnings and knowledge functions, and is beginning to play an effective role in societies across the globe [24]. Socio-cognitive robotics is the interdisciplinary study and application of robots that are able to teach, learn, and reason about how to behave in a complex world [24]. Social robotics technology promises many benefits but also challenges that society must be ready to confront with legal means and ethical imperatives.

Ethics describes moral principles that govern a person's or group's behavior. Roboethics captures the ethics and morals of robotics, the science of robots. Roboethics therefore integrates ethics into AI and algorithms. This field recently gained considerable attention among humanities and robotics engineers who draw on insights from computer science, artificial intelligence, mechanics, physics, math, electronics, cybernetics, automation, and control [24].

What specifies the emergence of socio-cognitive robotics is that humanity is at the threshold of replicating an intelligent and autonomous agent [24]. In order to enhance the ability of social robots to successfully operate in humane ways, roles, and environments, they are currently upgraded to a new level of physical skills and cognitive capabilities that embrace core social concepts [24]. Robotics thereby unifies two cultures, in which complex concepts—like learning, perception, decision-making, freedom, judgment, emotions, etc.—may not have the same semantic meaning for humans and machines [24].

In the organizational context, there is a divide projected into AI and non-AI humanness. Value may be derived from particular human traits, when many other manual and repetitive tasks will be outsourced to robots. Now we see that algorithms in finance are less likely preferred over human advisors, even if they are outperforming. In the organizational context, we already see a trend of reshoring formerly outsourced repetitive tasks to low-income countries back to the developed world with AI hubs. AI and automated control will generate economic superiority in the future, as humans having AI work for them can start using their additional time for creative human tasks. Leadership and organizational behavioral knowledge, which are features hard to replicate by AI, are expected to grow in the artificial age.

In the design and construction of social robots, the consideration of ethical concerns has therefore leveraged into an imperative [22]. Human–robot (a machine with a higher physical and social ability) interactions are somewhat different compared to other types of human–machine interactions (i.e., with a computer, cell phone, or other smart devices) [24, 32]. It is therefore essential for researchers, scholars, and users to clearly identify, understand, and consider these differences and ethical challenges so that they can benefit from and no one gets harmed by the assistance of social robots as a powerful tool in providing modern and quality services to society [24, 36].

Robots and algorithms now taking over human decision-making tasks and entering the workforce but also encroaching our private lives currently challenges legal systems around the globe [37]. The attribution of human legal codes to AI is one of the most groundbreaking contemporary legal and judicial innovations. Until now legal personhood has only been attached directly or indirectly to human entities [14]. The detachment of legal personhood from solely human beings now remains somewhat of a paradox causing an extent of "fuzziness" of the concept of personhood [6, 35, p. 1285]. As AI gets bestowed with quasi-human rights, defining factors of human personhood will need to be adjusted [14]. Human concepts, such as morality, ownership, profitability, and viability, will have a different meaning for AI. The need for redefining AIE has therefore reached unprecedented momentum.

While there is currently cutting-edge writing about the potential emergence of an AI personhood as well as concern over the merge of AI with cyberspace that might lead to the breach of the relationship between legal personhood and nation-state sovereignty and a nomenclature is emerging on legal characterizations of different levels of AI development, hardly any information exists about the eternal living of AI [19]. From the theoretical standpoint, the eternal longevity of AI contradicts the fundamental concept of fairness in death, as a general condition for all. From the practical standpoint, the international community is currently urged to think on the basis of global commons in terms of AI and AI eternal life potentials contributing to overpopulation. Thereby, global common theories may be tabbed on, which primarily offer guidance for a regulatory framework, which establishes control "… for the benefit of all nations" and refers to space constraints [11, 38].

Regarding limited space, longevity and eternal life of AI appear problematic. Humankind may face tough decisions whether or not to have AI proceed and what kind of developments to flourish and what to extinct. In what cases should we consider to switch off AI? In 1950, Isaac Asimov introduced the idea robot to (1) not injure a human being or, through inaction, allow a human being to come to harm; (2) a robot obeying the orders given it by human beings except where such orders conflict with the first law; and (3) a robot must protect its own existence as long as such protection does not conflict with the first or second law. In the cases of overpopulation and harm emerging from AI, algorithms and robots can be considered to be switched off. But when to stop AI?

Another killing market mechanism may be natural market selection via price mechanisms and the falling rate of profit. Regarding prices, natural supply and demand mechanisms will always favor innovation with a higher price and following supply of goods lead to a price drop. The falling rate of profit is one of the major underlying features of business cycles, long-term booms, and downturns [7, 8]. Capitalism is thereby described as competitive battle for innovation and reaping benefit from first-market introductions. Once followers enter the market, profit declines, leading eventually to market actors seeking novel ways to innovate in order to regain a competitive market advantage and higher rates of profit. Thereby, industries and innovations fade and die off. Such a natural market evolution is also likely to occur with AI innovations, which will determine which AI traits will remain and which ones will fade off.

Apart from soft market mechanisms that may lead to an AI evolution, what are the cases when AI should be shut down or switched off or—in the case if AI personhood—be killed? The main and leading concern about any new and emerging technology is to be safe and error-free [24]. Therefore, sufficient and numerous tests on health and safety must be performed by developers and/or well-known independent sources before rolling out any technology onto the marketplace and society [24]. In robotics, the safety issue mainly centers around software and/or hardware designs [24]. Even a tiny software flaw or a manufacturing defect in an intelligent machine, like a smart car or a social robot, could lead to fatal results [24]. When these deviations occur and especially when they are harmful to the human community but also to other AI species, the faulty AI should be terminated. With regard to the risk of robotic malfunctions and errors, product legal responsibility laws are mostly untested in robotics [24]. A usual way to minimize the risk of damage from social robots is to program them to obey predefined regulations or follow a code-of-ethics [24]. Ethical codes for robotics are currently needed and should become formed as a natural behavioral law to then be defined and codified as law. Laws but also an ethical understanding to terminate AI, algorithms, and robots in case of impairment and harm are needed.

As social robots become more intelligent and autonomous and exhibit enough of the features that typically define an individual person, it may be conceivable to assign them responsibility and use them in social, educational, and therapeutic settings [24]. In the currently ongoing research on the integration of computers and robotics with biological corpse, it is found that a cognizant human brain (and its physical body) apparently has human rights; hence, replacing parts of the brain with artificial ones, while not harming its function, preserves those rights [24, 40]. Also, consider a handicapped person featuring an electronic robot arm that commits a crime. It becomes obvious that half-robot–human beings should be considered as human and robots as quasi-human beings. Meghdari and Alemi [24] speculate that at some point in the future, we may face a situation in which more than half of the brain or body is artificial, making the organism more robotic than human, which consolidates the need of special robot rights and attributing (quasi)-human rights onto robots. When considering robots as quasi-human beings, their termination appears legally questionable and ethically challenging, requiring to revisit laws as legitimation to kill a likewise species as well as ethical consensus on the virtue of killing.

In its entirety, this book was the first introduction of AI ethics opening up many challenging questions in the behavioral context. For instance, what ethical code should we apply for controlling robots' actions? How can we program a switch to turn off AI in case of unlawful action and harm to people but also how to draw the boundary condition to ethical infringements? This is specifically important if humankind starts placing social robots in positions of authority, such as police, security guards, teachers, or any other government roles or offices, in which humans would be expected to follow them.

In the future, it is predicted that society is expected to fall into two extremes of a dichotomy between rationality (represented by AI) and humanness (represented by human beings). Hereby the question arises what is it that makes human humane? In the age of artificial intelligence and automated control, humanness is key to future success. Behavioral human decision-making insights and evolutionary economics can already today predict what makes human humane and how human decision-making is unique to set us apart from artificial intelligence rationality. Future research in these domains promise to hold novel insights on future success factors for human resource management but also invaluable contributions to artificial intelligence ethics.

Overall this book was meant as the first step toward a nomenclature of deciding on the future evolution grounded in the virtue of living and killing to motivate different viewpoints on the issue by cultural, religious, and ethical scholars. The book plays an important role in the evolution of an AI and human-mixed society in order to ground stability and social harmony into the newly emerging system. Depicting ethical imperatives around the life and death of machines being considered as quasi-human beings during this unprecedented time of societal change and regulatory reform holds invaluable historic opportunities for global governance policy-makers to snapshot the potential but also save from the likely downfalls of a robot–human mixed society.

The quantitative relationship between labor hours worked and the real prices charged for information should be expressed in probabilistic terms. In order to transform the value of information sharing into a price of a commodity, research needs to unravel the process whereby people give up some of their privacy for the sake of information sharing and how the products move into markets and are shared with nudgitalists. All these insights will aid to measure a hypothetical price according to pre-existing product values, which are socially-established prior to exchange. The transformation of commodity values into prices of production within nudgitalism and the creation of the surplus value or gross profit component of information are future research areas.

Currently, we lack understanding of how an output is produced, what utility is derived from social media that leads consumer-workers to give up privacy, and how exactly value is realized upon data amalgamation and sale in markets to nudgitalists. Social media advancement may be strongly affected by the sales income that social media producers get from selling data.

In the next research steps, a stringent hypothesis testing of the presented problem is recommended. A multi-faceted research plan to study the presented social deficiencies in a well-informed behavioral as well as heterodox way is needed. In such an approach, the book provides the first heterodox analysis of behavioral economics in the public sector.

Future research prospects should therefore aim at gaining insights of behavioral approaches to decision-making and develop critical analytical methods to spearhead behavioral economics applied in the public-sector domain. An introduction to the classic but also heterodox behavioral economics literature will aid in gaining a basic understanding but also critical reflection of contemporary behavioral insights attempts to tackle the most pressing societal concerns.

Future research projects featuring a multi-methodological approach will help derive invaluable information about the interaction of economic market with the real-world economy with direct implications for policy-makers alongside advancing an upcoming scientific field. Following empirical investigations should employ a critical survey of the intersection of analytic and behavioral perspectives to decision-making in the public sector. Literature discussion featuring a critical analysis of how to improve the behavioral economics approach to tackle critical public-sector challenges should be coupled with research training and enhancement of scientific skills. Research should be directed toward a critical analysis of the application of behavioral economics on public concerns.

In the behavioral economics domain, both approaches, studying the negative implications of imperfect behavior on judgment and decision-making of public servants but also finding ways how to train public servants making wiser decisions in leading citizens and pro-social choices should be explored.

Interdisciplinary viewpoints and multi-method research approaches should be covered in the heterodox economics readings but also in a variety of independent individual research projects. Research support and guidance should be targeted at nurturing interdisciplinary research interests in the fields of behavioral economics and public affairs. More concretely, future studies should define the value that data has to individuals and data sovereignty in the international context. When people share information, they should be informed to consider what the benefit and value of sharing are for them and what the benefit for social media industrialists-capitalists is. The sovereignty of data and the human dignity of privacy should become debated as virtual virtue in the twenty-first century. Individuals should be informed that sharing data is a personal security risk, if considered to be asked for social media information upon entry of a country.

Future studies should describe what companies and institutions constitute the complex system in order to help establish the nudgital society and the influence that social media has. Politicians may use various channels and instruments to manipulate the populace with targeted communication. The implicit underlying social structure of the nudgital society based on a complicated information gathering machinery should become subject to scrutiny and how, in particular, the nudgital class division is supported by a comprehensive social network data processing method. How social media advertising space can be used to specialize in targeted propaganda and misleading information to nudge the populace in an unfavorable way should be unraveled.

In previous US elections, the profit and value of detailed market information have been found to have gained unprecedented impetus. Future research should also draw a line between the results of the US presidential elections and the study of understanding heuristics to elucidate that heuristics played a key role for voting logically. This would be key as it would help explain how people chose to vote. This line of research could help to more accurately promote future elections' candidates, how to better predict election outcomes, and how to improve democracy.

Careers should also be examined to determine whether, in the course of the selection process, all people found fair competition online. In the future, it would be of crucial importance to clarify how people, in the face of misinformation or misrepresentation, allow themselves to be influenced by online search results and possibly make wrong or unfair decisions.

In addition, nudging through means of visual merchandising, marketing, and advertising should be captured in order to uphold ethical standards in social media. Nudging's role in selling products, maximizing profits but also creating political trends or influencing selection processes should be uncovered. While there is knowledge of the visual merchandising in stores and window displays, little appears to be known how online appearances can nudge people into making certain choices. In particular, the familiarity heuristic, anchoring, and the availability heuristic may play a role in implicitly guiding people's choices and discreetly persuading consumers and the populace. Targeted ads have become a prominent way to pick up interested consumers recently. Not to mention advancements of online shopping integrity and e-commerce ethics but foremost in order to curb harmful side effects of negative SEO, the prospective insights gained will aid uphold moral standards in economic marketplaces and hopefully improve democratic outcomes of voting choices but also enable more inclusive excellence in career placement decisions based on online contents.

Future research should also investigate how search engines can be manipulated to make favorable sources more relevant and how artificial intelligence and social networks can become dangerous data manipulation means. The role of data processing companies may be studied in relation to the idea of data monopoly advantages—hence, situations in which data processing companies may utilize data flows for their own purposes to support sponsored causes or their own ideals. All these results appear useful in determining future behaviors.

The current research in this area lacks empirical evidence, demanding further investigations on how nudges can directly impact individual's choices and new media can become a governance manipulation tool. What social instruments are employed on social media and what prospects data processing has in light of privacy infringement lawsuits should be uncovered. How social media is utilized to create more favorable social personas for political candidates should be explored. How Internet online presences allow to gain as much attraction as possible for the presence of political candidates is another question of concern. All these endeavors will help in outlining the existence of social media's influence in governance and data processing to aid political campaigning in order to derive inferences about democracy and political ethicality in the digital age.

Another area of concern is how selective representations influence the voting population and what institutions and online providers are enabling repetitiveness and selectivity. How gathered individual information is used to parse data to manipulate social Internet behavior and subsequent action is another topic to be investigated. Future research goals will include determining what this means for the future political landscape and how Internet users should react to political appearances online. Consumer education should target at educating social media users about their rights and responsibilities on how to guard their own and other's privacy.

How social media tools nudge people to not give everything at once but put it together in a novel way that it creates surplus should be analyzed. In small bits and pieces, individuals give up their privacy tranche by tranche. Small amounts of time are spent time on time. People, especially young people, may have a miscalibration about the value of information released about them. Based on hyperbolic discounting myopia, they may underestimate the total future consequences of their share of privacy.

Future studies should also look into the relationship between individual's political ideologies and how they use and interact on social media, especially with a focus on the concept of fake news and alternative facts. Where do these trends come from and who is more susceptible to these negative impacts of the digital society? Has social media become a tool to further polarize political camps needs to be asked. What implicit role have search engines gained in selection processes and how can search engine results be manipulated needs to be revealed in order to sensitize the general public about this growing trend as well as instigate collective action to combat unfair searchplace discrimination.

In addition, innovative means should be found to restore trust in media information and overcome obstacles such as the availability heuristic leading to disproportionate competitive advantages of media controlling parties. Information should be gathered on how we choose what media to watch and if political views play a role in media selection and retention. Does distrust in the media further political polarization and partisanship, needs to be clarified.

The preliminary results may be generalized for other user-generated web contents such as blogs, wikis, discussion forums, posts, chats, tweets, podcasts, pins, digital images, videos, audio files, and advertisements but also search engine data gathered or electronic devices (e.g., wearable technologies, mobile devices, 5G and the Internet of Things). Certain features of the nudgital society may also hold for tracking data, including GPS, geolocation data, traffic, and other transport sensor data and CCTV images, or even satellite and aerial imagery. All preliminary results should be taken into consideration for future studies in different countries to examine other cultural influences and their effects on social class and heuristics.

Promoting governance through algorism offers novel contributions to the broader data science and policy discussion. Future studies should also be concerned with data governance and collection as well as data storage and curation in the access and distribution of online databases and data streams of instant communication. The human decision-making and behavior of data sharing in regard to ownership should become subject to scrutiny in psychology. Ownership in the wake of voluntary personal information sharing and data provenance and expiration in the private and public sectors have to be legally justified [13].

In the future, institutional forms and regulatory tools for data governance should be legally clarified. Open, commercial, personal, and proprietary sources of information that get amalgamated for administrative purposes should be studied and their role in shaping democracy. In the future, we also need a clearer understanding of the human interaction with data and their social networks and clustering for communication results. The guarantee of safety of the information and the guarantee of the replacement or service, should a social media fail its function to uphold privacy law as intended, are additional areas of blatant future research demand. Novel qualitative and quantitative mixed methods featuring secondary data analysis, web mining, and predictive models should be tested for holding for the outlined features of the new economy alongside advancing randomized controlled trials, sentiment analysis, and smart contract technologies. Ethical considerations of machine learning and biologically-inspired models should be considered in theory and practice. Mobile applications of user communities should be scrutinized.

As for consumer–worker conditions, unionization of the social media workers could help uphold legal rights and ethical imperatives of privacy, security, and personal data protection. Data and algorithms should be studied by legal experts on licensing and ownership in the use of personal and proprietary data. Transparency, accountability, and participation in data processing should also become freed from social discrimination. Fairness-awareness programs in data mining and machine learning coupled with privacy-enhancing technologies should be introduced in security studies of the public sector. Public rights of free speech online in the dialogue based on trust should be emphasized in future educational programs. Policy implications of the presented ideas range from security to human rights and law to civic empowerment. Citizen empowerment should feature community efforts to protect data and information sharing to be free of ethical downfalls. Social media use education should be ingrained in standard curricula. Children should be raised with an honest awareness of their act of engagement on social media in the nudgital society of the digital century.

As for digital online inequality, the book offered a comparative Behavioral Law and Economics approach to capture the most contemporary trends in digitalization around the world with particular attention to inequality. The book started with a description of digitalization encroaching modern society and then outlined novel emerging inequalities in the wake of our current digitalization disruption.

The book made the empirical case for rising digitalization encroaching our societies around the globe. In historical overviews and backed by research results, readers were taken onto a journey to understand today's most recent advancements of digitalization with special attention to finance.

The book also shed light on a concurrent development of inequality as one of the most significantly-pressing concerns of our times within society, around the globe and over time. While the standard economic innovation literature assumes constant improvement of digitalization being available to the coming generations, the book also drew the heterodox case of a connection of innovation and inequality.

As for future research avenues, attempts to alleviate the impacts of inequality imbued in digitalization through law, economics and politics should be debriefed upon with a range of stakeholders. The role of the Internet in raising awareness, mobilizing e-social pressure and crowd control but also the negative aspects of online cancel cultures and the loss of classic media control in the age of digitalization should be discussed. Subsequent research could bring together digitalization and inequality in empirical evidence, theoretical advancements and vivid real-world relevant examples.

After introducing a working definition of digitalization inequality, the emergence and historical evolvement of inequality in innovations could be discussed. A historical snapshot of contemporary digitalization pressures would cover the aftermath of the 2008 World Financial Recession and our fading post-COVID-19 economy in regards to digitalization.

Future research could also considers rescue and recovery funding, a governmental tool currently applied to find financing in the private sector and use market incentives to lead public transitions to a more equitable world and, therefore, a more harmonious society. Most recent developments in 5G, cryptocurrencies, democratization of information and social media revenue repatriation as well as digital space exploration should become subject to scrutiny.

An analytic framework to dissect inequality in digitalization could be proposed comprised of qualitative and quantitative parameters in order to guide on a monitoring and evaluation agenda to see if the digitalization disruption can be delivered in an ethical and inclusive way. Behavioral Law and Economics rational but also disparate impact analysis tools should be presented in order to outline potential digital inequality alleviation strategies—such as education, skills development, institutional adjustment and favorable societal norms changing capacities. A comparative analysis of some of the most dominant digitalization hubs in Asia, Europe and North America would round up the picture of digitalization with a global outlook and extraterrestrial space exploration vision.

Paying homage to the innovative character of digitalization, future research endeavors should be dedicated to capturing the most recent developments in digitalization with particular attention to hidden inequalities. Societal divisions in access to digitalization but also global disparities are obvious inequalities on the surface. Below the obvious inequality, the greater public should be guided to discover digital competition in the age of online searchplace dominance.

Digital currencies and reclaiming economic growth and remuneration potential but also taxation revenue in the age of tangible and intangible online transactions could be discussed with the broad-based public. The role of digitalization in supporting COVID-19 prevention but also in integrating COVID Long Haulers back into the economy could be envisioned.

As the most wide-reaching and long-term innovation of the future, digital exploration of extraterrestrial territories could become thematized with raising attention to digital ethics in these novel colonization efforts. Lastly, to pay tribute to future generations, the contested relation between digitalization and sustainability should be reflected upon. Digitalization could be outlined as the remedy but also a burden in regards to the sustainable development goals opening up the final reflection upon potential digitalization inequality alleviation strategies.

Innovation could be heterodoxly-described as featuring problematic negative externalities. The human-imbued drive to constant change could thereby be dissected as the problem but also the solutions to combat digital inequality. Additional remedies discussed may include regulation (with particular emphasis on the Brussels effect), the economic impetus of taxation but also global governance institutions' role in monitoring and evaluating the concerted efforts to deliver a fair digitalized world.

A plaidoyer for Law and Economics analytics but also stakeholder management in the concerted action of a widespread digitalization disruption should be emphasized. The need for targeted Internet oversight agencies but also favorable behavioral social norms imbued in digitalization ethics will round up the future-oriented, interdisciplinary and wide-reaching digital ethics advancements.

Overall, the book will hopefully innovatively develop new interpretations, understandings, and concepts of behavioral management. In compiling scholarship and theories on risk mitigation coupled with the financial sector insights on how to curb harmful decision-making, the book is meant to help create a central reference point and resources on aggregate information on nudging and winking within markets and society. The mapping of the derived results may help guideline balanced approaches to implement nudging and winking through concerted public, private, and global efforts.

The monograph was targeted at elevating the importance of behavioral economics scholarship while deriving implications on how to improve individuals' lives, group interactions as well as societal harmony. Emphasizing areas where to apply ethics in the digital age and where to promote individual heuristic strategies will hopefully lead to practical implications for the individual but also private industry market actors and public policy sector officials.

Future outlooks include the question of whether artificial intelligence should be built to resemble human decision-making or deviate from human decisions for the sake of rationality. The question of whether humanness is key to future success in the digital era should be raised. Thereby, the basic rationale is that with artificial intelligence being used for every rational task, humanness, hence, irrational feelings, emotions, and creativity, which may not be replicable by artificial agents, are prospected to become more valuable. In artificial intelligence hubs, humanness may become a key qualification of success. In the international arena, artificial intelligence hubs may become local points, at which 24/7 robotics reap benefits from around the world, whereas artificial intelligence lacking areas will feature a workforce that requires rest and will fall back in terms of international trade. E-outsourcing is therefore prospected to increase the divide between developed and underdeveloped territories.

Current international law developments may also draw from the monograph in learning about the evolution of behavioral insights. Behavioral economics evolved out of the understanding that real-world behavior did not correspond with what economic rational choice theory would predict human would do. The growing body of knowledge on how decision makers actually choose was then used to give advice how to craft policies and design decision-making architectures in order to help people make more rational and more favorable choices over time and in large cooperative entities and groups. The current International Law Commission, as a body of the United Nations, has set out to discuss and evaluate the discrepancy between the law-in-books, which is laid out in legal codes, constitutions, treaties, etc., in contrast to the law-in-practice, which is the actual application and performance according to the law. This newly-opening field of Behavioral Law is a similar reality check of the practice and actual conduct of law-in-books with ample applications, for instance, as currently-scrutinized in the most pressing international customary law predicaments of rising sea levels and moving borders or refugee crises due to climate change. International law leaders around the world could draw from the insights gained in behavioral economics and behavioral insights for the current adjustment of laws and legal frameworks to fit human decision-making needs with respect for human fallibility and the behavioral tendencies.

In sum, understanding the different decision-making approaches but also shedding light on previously-unknown interdependencies of nudging and societal divides will aid to pursue establishing a more just humankind in the digital age. For practitioners, the results are esteemed to help in lowering institutional downfalls of increasingly interconnected and fragile global networks. For academia, the book may hopefully spearhead interdisciplinary research on behavioral economics and ethics in the artificial intelligence age. The link of nudging within the artificial intelligence revolution may help develop real-world-relevant public policy prescriptions for governments, private sector stakeholders as well as the individual. For the general public, legal aspects on privacy in the digital age as well as other resources on the coping strategies in light of libertarian paternalism but foremost practical advice how to use heuristics' potential to improve well-being will hold short-term innovative practical advantages as well as long-term invaluable historic assets on the blend or divide of human nature with or from artificial intelligence. It remains to be seen concretely how connectivity will affect general international development and economic growth in the digital age. As the Coronavirus pandemic crisis intensified digitization trends given changed consumer and social behavior enormously, it is already clear that we live in a rapidly-changing time of upheaval and that it is up to people to arise more resiliently, ethically strengthened and more humanely from these changes.

References

1. Alemi, M., Meghdari, A., & Saffari, E. (2017). RoMa: A hi-tech robotic mannequin for the fashion industry. In *Lecture Notes in Computer Science (LNCS): Social Robotics* (Vol. 10652, pp. 209–219). Springer.
2. Armstrong, S., & Sotala, K. (2012). How we're predicting AI—Or failing to. In J. Ramportl (Ed.), *Beyond AI: Artificial dreams* (pp. 52–65). University of West Bohemia.
3. Asimov, I. (1942/1950). *I, Robot*. Bantam Dell.
4. Asimov, I. (1978). My own view. In R. Holdstock (Ed.), *The encyclopedia of science fiction*. St. Martin's Press.
5. Asimov, I. (1985). *Robots and empire*. Doubleday.
6. Barrat, J. (2013). *Our final invention: Artificial intelligence and the end of the human era*. St. Martin's Press.
7. Brenner, R. (2002). *The boom and the bubble: The US in the world economy*. Verso.
8. Brenner, R. (2006). *The economics of global turbulence: The advanced capitalist economies from long boom to long downturn, 1945–2005*. Verso.
9. Căpek, K. (1921/2004). *Rossum's universal robots*. Penguin.
10. Cellan-Jones, R. (2014, December 2). Stephen Hawking warns artificial intelligence could end mankind. *BBC News*. www.bbc.com/news/technology-30290540
11. Clancy, E. (1998). The tragedy of the global commons. *Indiana Journal of Global Legal Studies, 5*(2), 601–619.
12. Copeland, J. (2000, May). What is artificial intelligence? *AlanTuring.net*. www.alanturing.net/turing_archive/pages/Reference%20Articles/what_is_AI/What%20is%20AI02.html
13. Donahue, J. D., & Zeckhauser, R. J. (2011). *Collaborative governance: Private sector roles for public goals in turbulent times*. Princeton University Press. https://doi.org/10.1515/9781400838103
14. Dowell, R. (2018). Fundamental protections for non-biological intelligences or: How we learn to stop worrying and love our robot Brethren. *Minnesota Journal of Law, Science & Technology, 19*(1), 305–336.
15. EU Committee on Legal Affairs. (2016, May 31). *Draft report with recommendations to the Commission on Civil Law Rules on Robotics*.
16. Galeon, D., & Reedy, C. (2017, October 5). Kurzweil claims that the singularity will happen by 2045. *Futurism*. www.futurism.com/kurzweil-claims-that-the-singularity-will-happen-by-2045/
17. Hayes, A. (2018a). *Decentralized banking: Monetary technocracy in the digital age* (Social Science Research Network Working Paper). https://papers.ssrn.com/sol3/papers.cfm?abstract_id=2807476
18. Hayes, A. (2018b). *The active construction of passive investors: Toward robo economicus* (Working Paper). University of Wisconsin-Madison, Department of Sociology.
19. Hildebrandt, M. (2013). Extraterritorial jurisdiction to enforce in cyberspace? *Toronto Law Journal, 63*, 196–224. https://doi.org/10.3138/utlj.1119
20. Kowert, W. (2017). The foreseeability of human-artificial intelligence interactions. *Texas Law Review, 96*, 181–204.
21. Larson, D. A. (2010). Artificial intelligence: Robots, avatars, and the demise of the human mediator. *Ohio State Journal on Dispute Resolution, 25*, 105–164.
22. Lin, P., Abney, K., & Bekey, G. A. (2012). *Robot ethics: The ethical and social implications of robotics*. The MIT Press.

23. Marra, W., & McNeil, S. (2013). Understanding "the loop": Regulating the next generation of war machines. *Harvard Journal of Law & Public Policy, 36*, 1139–1187.
24. Meghdari, A., & Alemi, M. (2018). Recent advances in social & cognitive robotics and imminent ethical challenges. In *Proceedings of the 10th International RAIS Conference on Social Sciences and Humanities Organized by Research Association for Interdisciplinary Studies (RAIS) at the Erdman Center at Princeton University*, Princeton, NJ. The Scientific Press.
25. Puaschunder, J. M. (2017a). Nudgital: Critique of behavioral political economy. *Archives of Business Research, 5*(9), 54–76.
26. Puaschunder, J. M. (2017b). Nudging in the digital big data era. *European Journal of Economics, Law and Politics, 4*(4), 18–23.
27. Puaschunder, J. M. (2017c). Nudgitize me! A behavioral finance approach to minimize losses and maximize profits from heuristics and biases. *International Journal of Management Excellence, 10*(2), 1241–1256. https://doi.org/10.17722/ijme.v10i2.957
28. Puaschunder, J. M. (2017d). The nudging divide in the digital big data era. *International Journal of Research in Business, Economics and Management, 4*(11–12), 49–53.
29. Puaschunder, J. M. (2018). *Artificial intelligence ethik* (Social Science Research Network Paper). https://papers.ssrn.com/sol3/papers.cfm?abstract_id=3137926
30. Puaschunder, J. M. (2019a). Artificial intelligence evolution: On the virtue of killing in the artificial age. *Journal of Sociology, 3*(1), 10–29.
31. Puaschunder, J. M. (2019b, June 10–11). Artificial intelligence market disruption. In *Proceedings of the International RAIS Conference on Social Sciences and Humanities Organized by Research Association for Interdisciplinary Studies (RAIS) at Johns Hopkins University*, Montgomery County Campus, Rockville, MD (pp. 1–8).
32. Saffari, E., Meghdari, A., Vazirnezhad, B., & Alemi, M. (2015). Ava (a social robot): Design and performance of a robotic hearing apparatus. In *LNCS: Social Robotics* (Vol. 9388, pp. 440–450). Springer.
33. Schuller, A. (2017). At the crossroads of control: The intersection of artificial intelligence in autonomous weapon systems with International Humanitarian Law. *Harvard National Security Journal, 8*, 379–425.
34. Sofge, E. (2015). Bill Gates fears A.I., but A.I. researchers know better. *Popular Science*. www.popsci.com/bill-gates-fears-ai-ai-researchers-know-better
35. Solum, L. (1992). Legal personhood for artificial intelligences. *North Carolina Law Review, 70*(4), 1231–1287.
36. Taheri, A. R., Meghdari, A., Alemi, M., & Pouretemad, H. R. (2018). Human-robot interaction in autism treatment: A case study on three pairs of autistic children as twins, siblings, and classmates. *International Journal of Social Robotics, 10*(1), 93–113. https://doi.org/10.1007/s12369-017-0433-8
37. Themistoklis, T. (2018). Artificial intelligence as global commons and the "international law supremacy" principle. In *Proceedings of the 10th International RAIS Conference on Social Sciences and Humanities Organized by Research Association for Interdisciplinary Studies (RAIS) at the Erdman Center at Princeton University*, Princeton, NJ. The Scientific Press.
38. Tsagourias, N. (2015). The legal status of cyberspace. In N. Tsagourias & R. Buchan (Eds.), *Research handbook international law and cyberspace* (pp. 13–29). Edward Elgar Publishing.
39. United Nations Department of Economic and Social Affairs. (2017). *Will robots and AI cause mass unemployment? Not necessarily, but they do bring other threats*. https://www.un.org/development/desa/en/news/policy/will-robots-and-ai-cause-mass-unemployment-not-necessarily-but-they-do-bring-other-threats.html
40. Warwick, K., & Shah, H. (2014). How good robots will enhance human life. In K. Tchon & W. W. Gasparski (Eds.), *Treatise on good robots edition: Praxology: The international annual of practical philosophy and methodology* (pp. 3–19). Transaction.
41. Wisskirchen, G., Biacabe, B. T., Bormann, U., Muntz, A., Niehaus, G., Jiménez-Soler, G., et al. (2017). *Artificial intelligence and robotics and their impact on the workplace*. IBA Global Employment Institute.

The manufacturer's authorised representative in the EU is Springer Nature Customer Service Centre GmbH, Europaplatz 3, 69115 Heidelberg, Germany. If you have any concerns regarding our products, please contact ProductSafety@springernature.com

Printed and bound by CPI Group (UK) Ltd, Croydon, CR0 4YY

30/09/2026

02225228-0007